Water Rights: An International and Comparative Study

Water Law and Policy Series

The role of law in the development and management of the world's water resources is often overlooked by water resource experts. This series aims to make "water law" - international, national and transnational - more accessible to the global water community.

While focusing primarily on the legal component of water resources management, the series will integrate other water-related disciplines in order to ensure a truly comprehensive and interdisciplinary approach to the problem.

SERIES EDITORS

Patricia Wouters
International Water Law Research Institute, University of Dundee, UK.

Sergei Vinogradov
Centre for Energy, Petroleum and Mineral Law and Policy,
University of Dundee, UK.

Other titles in the series include:

A New Legal Framework for Managing the World's Shared Groundwaters: *A Case-study from the Middle East*

International Law and Sustainable Development: *Lessons from the Law of International Watercourses*

Hydrology and Water Law – Bridging the Gap

Water Rights: An International and Comparative Study

Desheng Hu
Professor of Law, Zhengzhou University, China

IWA Publishing
London • Seattle

Published by IWA Publishing, Alliance House, 12 Caxton Street, London SW1H 0QS, UK
Telephone: +44 (0) 20 7654 5500; Fax: +44 (0) 20 7654 5555; Email: publications@iwap.co.uk
Web: www.iwapublishing.com

First published 2006

British Library Cataloguing in Publication Data
A CIP catalogue record for this book is available from the British Library

Library of Congress Cataloging-in-Publication Data
A catalog record for this book is available from the Library of Congress

ISBN 1843391082
ISBN13 9781843391081

Contents

Preface

"In this new century, water, its sanitation, and its equitable distribution, pose great social challenges for our world. We need to safeguard the global supply of healthy water and to ensure that everyone has access to it. ... Access to safe water is a fundamental human need and therefore a basic human right"

- Kofi Annan, United Nations Secretary-General [1]

The complex issue of "water rights" is an important topic that is increasingly finding its way into global debate and discourse. The concept has been, and continues to be, used rather indiscriminately across the spectrum of water-related issues, usually within discussions surrounding "water as a human right". The matter has been examined recently within the context of a UN General Comment[2] and studied by various commentators,[3] with significant focus at the

1 http://www.righttowater.org.uk/.
2 See General Comment No. 15 on the Right to Water was adopted by the UN Committee on Economic, Social and Cultural Rights at its twenty-ninth session in November 2002 (UN Doc. E/C.12/2002/11).

recent 4th World Water Forum convened in Mexico.[4] The Ministerial Declaration issued at Mexico failed to satisfy a number of States[5] and led some to observe the lack of international commitment around the "right to water.[6] And the debate continues, without clear lines of implementation, in an area where the law is unclear.

Dr Hu's book provides new insight into how "water rights" might be understood at international and domestic level, and presents a most challenging case study – China – through which he systematically analyses the concept. Dr Hu has developed a unique analytical framework for examining rigorously the evolution of the notion of "water rights" and uses a three-dimensional prism to study the contours of the concept – "the property right", "the human right" and the "environmental right" to water. Each of these is examined in turn, with a focus on distilling out the normative legal aspects of each, which then form the foundation for devising a new water rights system.

Dr Hu presents eight basic principles for building this new legal regime, designed to meet the compelling water-related issues in the existing and near future. As one of the fastest growing economies in the world, China faces serious problems over the management of its water resources, with demand exceeding supply in many urban centres. The situation is exacerbated by global warming, population growth, demographic movements and gaps in the legal and regulatory system governing water. Indeed, many question how China will respond to these challenges.

---

[3]See inter alia, IUCN, "Water as a Human Right", J. Scanlon et al, IUCN; WMO, "The Right to Water", WMO (2003). See also http://web.amnesty.org/pages/ec-water-eng, where Amnesty International suggests that "governments have duties" to fulfil a "right to water", especially in times of scarcity.

[4] A summary of the water rights discourse at Mexico can be found at http://www.iisd.ca/ymb/worldwater4/html/ymbvol82num15e.html. It summarises that water rights was discussed under three thematic sessions: "the human right to water"; the civil society perspectives on securing the right to water; and "the normative meaning and issues surrounding implementation of the "right to water".

[5] Bolivia's Minister of Water refused to sign the Ministerial Declaration, claiming that it fell far short of ensuring a right to water; see "Bolivia Says Water is a Right, Opposes Declaration", Reuters, 22 March 2006, at http://www.ecoearth.info/articles/reader.asp?linkid=54242. This position seems to be spreading across Latin America with Cuba, Uruguay, and Venezuela presenting an alternative declaration affirming the human right to water.

[6] Right to Water, WaterAid http://www.righttowater.org.uk/.

A review of the existing China water law reveals some progress has been made, but reveals great room for improvement.[7]

This work is a welcome contribution to an area of law that requires more thought and reflection. Tackling the issue of "water rights" in China is an ambitious task – but one that Dr Hu has done admirably.

It is hoped he will continue his work on the topic in China where he now works and pioneer new approaches to the difficult legal issues surrounding "water rights".

Dr Patricia Wouters
Series Editor
Dundee, 4 May 2006

Director, International Water Law Research Institute
UNESCO Centre for Water Law, Policy and Science
University of Dundee, Dundee, Scotland
Email: p.k.wouters@dundee.ac.uk

7 P. K. Wouters et al, "The New Development of Water Law in China", 7 University of Denver Water Law Review, pp.243-308, (2004).

Abbreviations

CDWA	Competent department of water administration
CESCR	Committee on Economic, Social and Cultural Rights
CGM	Municipality directly under the Central Government
China Water Law 1988	Water Law of the PRC 1988
China Water Law 2002	Water Law of the PRC (2002 Revision)
COAG	The Council of Australian Governments
CPC	Communist Party of China
DWAF	Department for Water Affairs and Forestry of South Africa
DWR	Department for Water Resources of South Australia
ECEL	European Council of Environmental Law
FAO	Food and Agricultural Organization of UN
IUCN	IUCN - The World Conservation Union
MWR	Ministry of Water Resources of the PRC
NDRC	National Development and Reform Commission
NPC	National People's Congress of the PRC
NWRS	National Water Resource Strategy
OECD	Organisation for Economic Co-operation and Development
PRC	People's Republic of China
UK	United Kingdom of Great Britain and Northern Ireland

UN	United Nations
UNDP	United Nations Development Program
UNECE	United Nations Economic Commission for Europe
UNESCO	United Nations Educational, Scientific and Cultural Organization
USA	United States of America
USGS	United States Geological Survey
USSR	Union of Soviet Socialist Republics
WHO	World Health Organisation
WMA	water management area
WMO	World Meteorological Organisation

for my parents

1

Introduction

The world's water resources are in crisis, in terms of water scarcity and quality as well as water-related disasters. Furthermore, many interdependent changes in the global context, such as geopolitical changes, population growth, agricultural demand, energy requirements, urbanisation, economic growth and industry, globalisation, technological changes, lifestyle, recreation and tourism, plus climate change, have been continuing to affect the water situation, and water stress looks set to augment radically in 60 percent of the world. In order to combat this water crisis, it is indispensable to improve the management of water resources.[1] As a vital natural resource, water is essential for sustaining all forms of life and conserving a beneficial environment, as well as for social and economic development. Water resources are not only a social good but also an economic good. Any limited natural resource needs effective and efficient administration or management, and water resources should not be excluded

[1] UNESCO, et al. (2003) *The United Nations World Water Development Report: Water for People, Water for Life*, UNESCO Publishing/Berghahn Books, Paris. pp. 4-21.

ISBN: 1843391082. Published by IWA Publishing, London, UK.

from this.[2] Either from the viewpoint of a lawyer, or from that of a hydrologist, or even from the viewpoint of an expert in integrated water resources management, water law should play an important role in the development and utilisation of water resources.[3] "Stability of water rights is an important principle in water law", and "[a] system of stable water rights is an incentive to invest in the development and conservation of water resources".[4] Where a water crisis exists, especially in a water-shortage jurisdiction, a legal mechanism for water rights should be well established or strengthened to ensure the sustainable, reasonable, equitable and efficient utilisation of water resources.

The central question intended for this work is: what principle(s)[5] and structure should be employed in designing an ideal water rights system for, or

---

[2] These views have been recognized in many international documents, *e.g.*, *The Dublin Statement on Water and Sustainable Development* (hereinafter "*The Dublin Statement on Water and Sustainable Development 1992*"), adopted by International Conference on Water and the Environment 1992, UN Doc.A/CONF.151/PC/112; Chapter 18 of *Agenda 21: A Programme for Action for Sustainable Development* (hereinafter "*Agenda 21*"), Rio de Janeiro, Brazil, Jun. 13, 1992, in *Report of the United Nations Conference on Environment and Development*, Annex II, U.N. Doc. A/Conf.151/26 (Vol. II) (1992); *Ministerial Declaration of The Hague on Water Security in the 21st Century 2000*, available at http://www.gdrc.org/wem/water/hague-declaration.html (last visited 31/05/2005).

[3] See Wouters, P. K., et al. (2001) The Legal Response to the World's Crisis: What Legacy from the Hague? *Water Law Review* 4, 418. Linsley, R.K., et al. (1992), *Water-Resources Engineering* (4th ed.) McGraw-Hill, London, p. 169. Gleick, P. H., et al. (1998) *The World's Water 1998-1999*, Island Press, Washington, D.C., p. 155.

[4] Solanes, M. and Gonzalez-Villarreal, F. (1999) The Dublin Principles for Water as Reflected in a Comparative Assessment of Institutional and Legal Arrangements for Integrated Water Resources Management, Global Water Partnership, Stockholm, p. 29.

[5] The term "principle" used here refers to legal principle. The word principle has many different meanings. For example, eleven meanings are given in The Oxford English Dictionary, in which 5 a) is more relevant to law science. A principle is "a fundamental truth or proposition, on which many others depend; a primary truth comprehending, or forming the basis of, various subordinate truths; a general statement or tenet forming the (or a) ground of, or held to be essential to, a system of thought or belief; a fundamental assumption forming the basis of a chain of reason". *The Oxford English Dictionary* (2ed.) (Vol. XII), 1989, Clarendon Press, Oxford, pp. 498-500.

In Bouviers Law Dictionary (1856 ed.), the term "principles" is defined as, "[b]y this term is understood truths or propositions so clear that they cannot be proved nor contradicted, unless by propositions which are still clearer. They are of two kinds, one when the principle is universal, and these are known as axioms or maxims; as, no one can transmit rights which he has not; the accessory follows the principal, &c. The other class are simply called first principles. These principles have known

improving an existing one in, a water-shortage jurisdiction? Related to this, are the following sub-questions: what should be the object(s) of water rights? what should be the legal format and substantive content of water rights? how and to what extent should the different aspects of water rights be integrated into the whole domestic legal system? and how should the government administer water rights so that they are properly implemented and well protected?[6] For any water-shortage jurisdiction, the above research questions need to be addressed very carefully. Only then, can a water-shortage jurisdiction succeed in either establishing a solid new water rights system, or improving and perfect an existing one.

Two kinds of methodology, both an analytical and a comparative approach, will be employed as the main basis for this work, with a focus on the People's Republic of China (PRC).

Why focus on China? China has experienced serious water crises in the past, and remains threatened by the increased potential for future water-related catastrophes, including natural disasters such as drought and flooding.[7] In order to address these issues, in October 2000, the Ministry of Water Resources (MWR) of the PRC introduced the concepts of "water rights and water market". Since that time, the issues relating to water rights, broadly defined, have become hot topics for discussion in both the public and private arenas, and many complex problems have been raised. The main concerns arise from the imprecise definition for "water resource", and the widespread misunderstanding of the term "water rights". This situation has confused the public as well as the

---

marks by which they may always be recognized. These are, 1. That they are so clear that they cannot be proved by anterior and more manifest truths. 2, That they are almost universally received. 3. That they are so strongly impressed on our minds that we conform ourselves to them, whatever may be our avowed opinions." See the explanation of "PRINCIPLES" in *Bouviers Law Dictionary* (1856 ed.), available at http://www.jusbelli.com/Bouvier/bouvier1856_p_all.html (last visited at 31/05/2005).

The author would like to employ the term "principle" as meaning a fundamental to general truth, or a comprehensive law or doctrine, from which others are derived, or on which others are founded. Therefore, a principle "differs in three ways from a rule: (a) Rules operate in an all or nothing fashion. If a case falls within the conditions of application for a rule, then the rule must be applied (*i.e.* it is conclusive). Principles, by contrast, have the quality of 'weight' or importance. (b) Rules specify a course of action to be followed or avoided. Principles, by contrast, are about *values* and cannot, therefore, indicate particular actions. (c) Rules can be justified by principles, but the reverse is not true." See Wilkinson, D. (2002) *Environment and Law*, Routledge, New York, p. 100.

[6] The reader should note that, this work would not conduct the research on how to deal with the existing *de facto* right to draw water in a new recommended water rights system for China.

[7] See the detailed discussion in 3.1 of Chapter 3.

representatives of the National People's Congress (NPC) of the PRC.[8] An example of this is the fact that the recommendations concerning the transfers of water-drawing rights, as contained in the Water Law of the PRC (Amendment Draft), was not encompassed in the Water Law of the PRC (2002 Revision) (hereinafter "China Water Law 2002") adopted on 29th August 2002.[9] The confusion over the definition of the term "water rights" in law and practice has led to the utilisation of water resources in China in ways that are not sustainable, reasonable, or equitable. To resolve the current crisis and impending future ones, it is critical for China to establish an appropriate legal structure for water rights that should include scientific substantial and procedural stipulations.

An analytical approach will be employed to identify "water rights" issues in China and, in particular, examine the normative content of the property rights aspects of water resources management. This research will analyse "property rights" for the following reasons. Firstly, the realisation and protection of the human right to water and the environmental right to water require an effective and enforceable "property rights" based legal approach. Secondly, while experience shows that economic instruments can be effective tool(s) in the development and utilisation of water resources, it is submitted here that such an approach is best regulated by a property right based legal system.

After the problems relating to water rights in China are identified, a comparative analytical approach will be applied to this research. Firstly,

---

8 See Hu, D. (2004) Research on the Delimitation of Property Rights on Water Resources, *Journal of Henan Finance & Taxation College* **18** (1), 34-37, p. 50.

9 Article 45(2) of the Water Law of the People's Republic of China (Amendment Draft), which was proposed by the State Council on 26 November 2001, provided that: "[a]ny unit or individual which has obtained the water-drawing right, and has saved water . . . within the duration and quantity of the relevant licence, are entitled to transfer the saved water resources after the approval of the authorities which issued the relevant permit . . . .". See *Water Law of the People's Republic of China (Amendment Draft)*, attached in Guohun [2001] 156, dated 26 November 2001; also available at http://www.npcnews.com.cn/gb/paper12/1/class001200078/hwz211659.htm (last visited 31/05/2005).

However, some representatives thought that, the time for enacting the transfer of the right to draw water has not come. See Qiao, X. (2002) Report of the Law Commission of the NPC on the Revision of the Water Law of the People's Republic of China (Amendment Draft) (26 April 2002), Chinese version, *Gazette of the Standing Committee of the NPC* (2002) 5, 375–378. Finally, there is no provision concerning the transfer of water rights in the Water Law of the PRC (2002 Revision), which was adopted at the 29th Meeting of the Standing Committee of the Ninth NPC, and came into force on 1st October 2002. Water Law of the PRC (2002 Revision), official Chinese version, *Gazette of the Standing Committee of the NPC* (2002) 5, 362–370.

international law will be examined to determine whether lessons can be derived from current trends that might support the design of an effective domestic water rights legal system for a water-shortage jurisdiction, especially for China. Secondly, the work will examine how the relevant problems identified have been dealt with under the legal regimes of the selected domestic case studies.

Why international law? The increased population, economic development and the issues of human rights, environment and natural resources (including water resources) have become topics of global concern, especially since the 1970's. During the past two decades, internationalisation or globalisation in law and policy has become an important factor in the development of national legislation and policy-making. For example, under the key concept of sustainable development, although it represents a commitment to economic growth, "[a] central strand of policy at Rio was the emphasis on the conservation of natural resources".[10] The majority of countries, including China, have enacted their own programmes in accordance with Agenda 21.[11] In Chapter 18 of Agenda 21, which is entitled "Protection of the Quality and Supply of Freshwater Resources: Application of Integrated Approaches to the Development, Management and Use of Water Resources", the importance of freshwater resources is clearly emphasized.[12] Despite the fact that water law continues to be constructed and enforced within a national framework, the substantive rules contained in national water law are being increasingly influenced by international documents at some level. Accordingly, some of the mechanisms invoked in modern international law, especially those employed in international legal documents, *e.g.*, treaties, conventions, United Nations (UN) declarations and resolutions, etc., have influenced and should continue to influence the adoption of water policy and/or law in China.[13] However, the water rights discussed in this work do not directly involve the States' international obligations relating to international watercourses.

The second area of comparative analyses is the other selected national water law regimes. It should be appreciated, however, that no single other jurisdiction's legal regime for water rights could completely solve the water rights problems in China. In other words, it would be impossible to transplant a foreign approach for water rights into China. Therefore, this work will compare only certain features of the selected legal regimes, with a focus on those particularly relating to the specific problems in China. In establishing the

---

10 See Bell, S. & McGillivray, D. (2000) *Environmental law* (5th ed.), Blackstone Press, London, p. 16.

11 See, *National Implementation of Agenda 21 – 2002 Country Profiles*, available at UN webpage http://www.un.org/esa/sustdev/natlinfo/cp2002.htm (last visited 31/05/2005).

12 See Chapter 18, *Agenda 21*.

13 See the detailed discussion in 4.1 of Chapter 4.

criteria for the selection, the following factors have been taken into consideration: (a) evidence of problems with severe water shortages or that a crisis of water resources exists, and that the situation is more serious than in China; (b) inter-basin water transfer exists and is needed; (c) the water law has been reformed or modernised in the past decade in accordance with the relevant rules of modern international law; and (d) relevant achievement(s) have been made in the area relating to water rights. Using these criteria, South Africa and South Australia are chosen as the most relevant national jurisdictions for examination. Their water law systems related to water rights will be compared with those of China.

Why South Africa? In South Africa, water is a critical but scarce natural resource, distributed unevenly in time (frequent droughts alternate with periods of good rainfall) and space (the eastern half of the country is markedly wetter than the western half); its water crisis, particularly its water shortage, is more critical than that in China.[14] The increasing demand for water and decreasing water quality make careful water management a priority in South Africa. The situation is similar to that in China to some extent. A more important factor is that the present water law of South Africa emphasises the "human right" to water, including such concepts as access to basic services as a human right, equity, public or stakeholder participation and the environmental right to water. It is well known that the discriminatory laws and practices in South Africa before the first democratic election in April 1994 prevented equal access to water and the use of water resources. Although there is no similar background, the human rights aspect is not fully reflected in Chinese water law, and this work aims to analyse more closely the foundation for the human right to water as developed in South Africa.

Why South Australia? A lower and unevenly distributed precipitation, together with a high evaporation rates, results in a low recharge rates for surface water and groundwater by rainfall in Australia, and further makes Australia the driest inhabited continent in the world[15]—thus representing a suitable case study with a severe water shortage. To tackle the issues relating to water resources, with the implementation of the Water Reform Framework 1994 and the subsequent Future Information on the National Water Initiative 2004,[16] on the

---

[14] See the detailed discussion in 5.2.1 of Chapter 5.

[15] See the detailed discussion in 5.3.1 of Chapter 5.

[16] In its meeting in Hobart on 25 February 1994, the Council of Australian Governments (COAG) endorsed the Water Resources Policy and agreed to its implementation within a five to seven years period. The Water Resources Policy, which is usually called and well known as *Water Reform Framework 1994*, is the ATTACHMENT A of *Council of Australian Governments Communiqué (Hobart, 25 February 1994)*. COAG, *Water Reform Framework 1994*, also available at

one hand, a mechanism for the transfer of the right to take water has developed in Australia; whilst on the other, protection for hydro-ecology has been strengthened there.[17] However, under the federal system of government in Australia, the six States and the two Territories have the primary responsibility for natural resources management. For the purpose of this study, one particular state or territory should be selected. South Australia is selected, specifically because of its focus on the "property right" aspect of water resources and the water right of water-dependent ecosystems.[18] These elements need to be addressed in China and thus this study will analyse the approach adopted in South Australia, from which lessons may be learned.

In considering the establishment of a revised water rights system in China, the author believes that three key aspects should be taken in account: (a) the property right aspect of water resources, (b) the human right aspect of access to water, and (c) the environmental right aspect of water resources management. These key elements will be examined separately or jointly when discussing the preferred approach to the administration of the China's water rights system. As to the aspect of the "property right" on water resources, this will be discussed on the basis of the civil law system, which is the one adopted by the modern Chinese legal system.[19] However, there may still be useful lessons to be learned from certain methods employed under the common law jurisdictions, as may be gleaned from the South Australia water rights legal regime.

The objectives and/or contributions that are intended for this research are:

---

http://www.disr.gov.au/science/pmsec/14meet/inwater/app3form.html (last visited 31/05/2005).

On 29 August 2003, in order "to increase the productivity and efficiency of water use, sustain rural and urban communities, and to ensure the health of river and groundwater systems", COAG agreed to refresh Water Reform Framework 1994, and a Proposed National Water Initiative was recommended; see *Council of Australian Governments Communiqué (29 August 2003)*, available at http://www.ministers.dotars.gov.au/ja/releases/2003/august/a100_2003_attach.htm (last visited 31/05/2005). Later, on 25 June 2004, COAG approved the *Future Information on the National Water Initiative* in its meeting held in Canberra; see *Council of Australian Governments' Meeting, 25 June 2004, Canberra, Communiqué*, available at http://www.pm.gov.au/news/media_releases/media_Release941.html (last visited 31/05/2005), and the *Future Information on the National Water Initiative* is the Attachment A of the *Communiqué*.

[17] See COAG, *Water Reform Framework 1994*; also the latest assessment of progress in its implementation, *i.e.*, *2002 National Competition Policy Assessment Framework for Water Reform*, available at http://www.ucc.gov.aupdf/AST4AfrWa-001.pdf (last visited 31/05/2005).

[18] See the detailed discussion in 5.1 and 5.3 of Chapter 5.

[19] See the detailed discussion in 3.2 of Chapter 3.

(1) To identify the legal issues affecting the definition and administration of water rights in China; to accomplish this, three aspects will be considered, namely, the property right on water resources, the human right to water, and the environmental right to water;

(2) To select and/or analyse the regimes selected at the international and national levels, using comparative analysis;

(3) To seek possible solutions to the problems related to the management of water resources in water-shortage jurisdictions, and to make recommendations for a revised national water legal regime in China, including those who will hold and who will administer water rights.

In structure, this work consists of seven chapters. Following this introductory chapter, Chapter 2, "Water Resources and Water Rights: an Overview", will firstly provide a literature review about the key elements of "water", "water resource" and "water utilisation". Then, the evolution of water rights - a changing concept – will be analysed. Finally, the three aspects of water rights used for analyses– the property right on water resources, the human right to water and the environmental right to water – will be discussed in general terms.Following the analytical approach developed in Chapter 2, after basic information concerning the situation of water resources in China and modern Chinese legal system have been given, Chapter 3 "Water Rights in China: Law, Practice and Problems" will analyse water rights in law and practice, identify relevant issues relating to water rights in China. Chapter 4 "The Influence of International Law" will firstly identify the legal status of the human right to water and the environmental right to water. Then, their contents and the States' obligation in international law will be examined. Further, in Chapter 5 "Water Rights Systems in South Africa and South Australia", the legal systems governing water rights in South Africa, as well as in South Australia water law, will be analysed.

Based on the analyses presented in the previous four chapters, in Chapter 6 "Comments and Observations on Water Rights System", the author offers suggestions about the principles and structure for an ideal water rights system, and focuses on making recommendations to be adopted by China in order to improve its national legal regime governing water resources. Finally, in the last chapter, brief conclusions will be drawn up with a short summary that explains the basis for the recommendations set forth.

2

Water Resources and Water Rights: An Overview

Using a literature approach, this chapter will provide an overview of the term "water resource" and its "utilisation", and examine also the evolution of water rights in a broader sense. Unless the concepts of "water resource" and its "utilisation" are well understood, it is difficult to discuss water rights competently. The first part of this chapter examines, in some detail, the meaning of these terms. Following this, this work reviews the evolution of the notions concerning water rights. Finally, the three aspects of water rights, *i.e.*, the property right on water resources, the human right to water, and the environmental right to water, are examined in general terms. On the whole, this chapter provides the foundation for the ensuing discussions and analyses.

2.1 WATER RESOURCES AND ITS UTILISATION

Water resources and/or its utilisation are the objects of water rights, and play a key role in influencing the types, as well as the legal content of water rights. A

ISBN: 1843391082. Published by IWA Publishing, London, UK.

sound understanding of water resources and its utilisation is a necessary foundation for an examination of water rights. After these have been discussed through a literature review, relevant legal issues will then be explored.

2.1.1 Definition of water resource

This section discusses the definitions of water resource beyond the legal perspective through a literature review. A full consideration of these definitions is useful in determining the legal definition of water resource in, or suggesting an appropriate one for, a particular domestic water legal regime.

2.1.1.1 Water properties[1]

Recognising the chemical and physical properties of water facilitate a clearer understanding of why water exists in many places in different states, and why water can play multiple functions. Using a literature review, the following will be examined.

The chemical properties of water: The chemical description for water is H_2O. With a positive charge on the one side (where the hydrogen atoms are) and a negative charge on the other (where the oxygen atom is), water molecules tend to attract each other. Water can diffuse the molecules of more things than any other liquid. This is to say, when water flows either through the ground or through bodies, it can remove with it numerous kinds of chemicals, minerals, nutrients, etc.

The physical properties of water: Water is the only natural thing found in all three states, *i.e.*, liquid, solid (ice, snow) and gaseous (steam, water vapour), at the temperatures usually found on the earth, and it is, therefore, unique. With a particularly high heat index, water can also absorb a large amount of heat before it starts to become hot.

2.1.1.2 Hydrologic cycle

The hydrologic cycle, also called the water cycle, is a dynamic, unending and continuous process through which water passes under natural or man-made conditions. Only after understanding the hydrologic cycle does it become clear

---

[1] This part is summarized from *Encyclopaedia Britannica* (1973), Vol. 23, pp. 271-271; Mazor, E. (1991) *Applied Chemical and Isotopic Groundwater Hydrology*, Open University Press, Milton Keynes, pp. 1-2; Beck, R. E., ed. (1991) *Water and Water Rights*, Vol. I., Charlottesville, p. 1; and United States Geological Survey (USGS), *Water properties*, available at USGS webpage http://ga.water.usgs.gov/edu/waterproperties.html (last visited 31/05/2005).

that it is unwise for a domestic legal system to manage all water in the hydrologic cycle.

The United Nations Educational, Scientific and Cultural Organization (UNESCO)/World Meteorological Organisation (WMO) defines the hydrologic cycle as a "[s]uccession of stages through which water passes from the atmosphere to the earth and returns to the atmosphere: evaporation from the land or sea or inland water, condensation to form clouds, precipitation, accumulation in the soil or in bodies of water, and re-evaporation".[2] Figure 2.1 demonstrates the hydrologic cycle, the movement of water and the three physical forms of water.

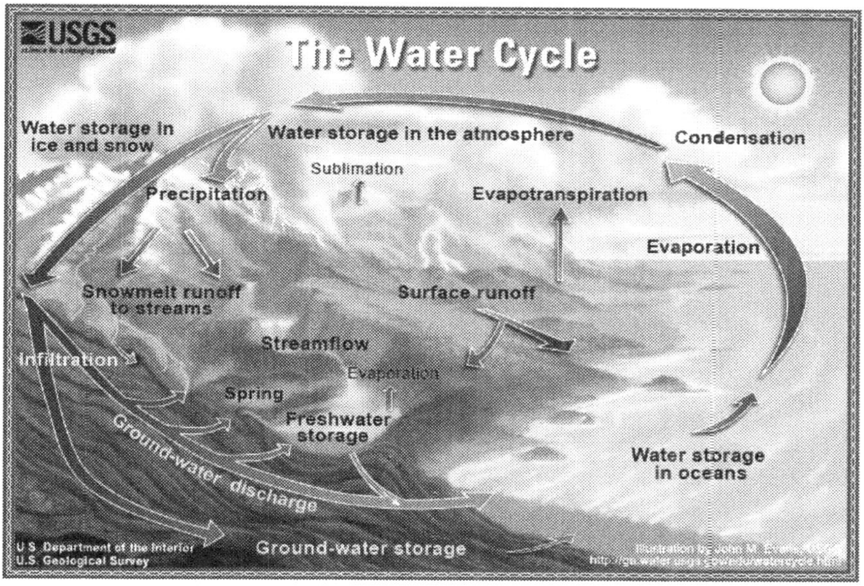

Figure 2.1. Diagram of the Water Cycle (USGS webpage http://ga.water.usgs.gov/edu/watercyclehi.html).

2.1.1.3 *Water quantity and location*

With the knowledge of how much water there is in/on the earth and where that water is, one may understand why the water rights system mainly regulates freshwater. This section will discuss the quantity and location of water in terms of the literature review.

---

2 UNESCO/WMO (1992) *The International Glossary of Hydrology* (2nd ed.), Paris. Its electronic version available at http://www.cig.ensmp.fr/~hubert/glu/aglo.htm (last visited 31/05/2005).

Although water is described as "one of the most abundant, widely distributed, and essential substances on earth",[3] freshwater only accounts for about 2.76 percent of the whole,[4] further, only less than one percent of all freshwater is available to be accessed and used by human beings.[5] When we talk about water, we usually mean freshwater, and that will be the major approach adopted in this work. Generally, freshwater is divided into three categories according to its location, *i.e.*, surface water – "[w]ater which flows over or is stored on the ground surface", groundwater – "[s]ubsurface water occupying the saturated zone", and atmospheric water – "[w]ater as it occurs in various forms in the air".[6]

2.1.1.4 Water resource: definitions from non-legal perspective

Understanding the definitions of "water resource" from the non-legal perspective is a useful basis for proposing an appropriate definition of "water resource" for water rights system from the legal viewpoint. Thus, a literature review approach is employed here to discuss these definitions.

Although a definite scope of water resource is indispensable for the sustainable integrated management of water resources, there is no universally accepted definition for "water resource". As to the meanings of "water resource", different definitions have been offered from different viewpoints, perspectives and/or approaches.

As to the word "resources", there exist various viewpoints.[7] In this work, the author would like to define "resources" from a non-anthropocentric perspective,

---

[3] See *Encyclopaedia Britannica* (1973), Vol. 23, p. 270.

[4] See USGS, *Where is Earth's water located?* available at http://ga.water.usgs.gov/edu/earthwherewater.html (last visited 31/05/2005).

[5] See UN, *Brochure on International Year of Freshwater*, available at http://www.un.org/events/water/brochure.htm (last visited 31/05/2005).

[6] See UNESCO/WMO, *supra* note 2.

[7] For example, from an anthropocentric viewpoint, Montgomery, C. W. defines it as "[i]n a general sense, resources are all those things that are necessary or important to human life and civilization, that have some value to individuals and/or to society. Implicit in the term is the availability for a supply that can be drawn on as required in the future. The requirement that resources have some value to people means that what constitutes a resource may change with time or social context. Many of the fuels, building materials, metals, and other substances important to modern technological civilization (which are thus resources in a modern content) were of no use to the earliest cave dwellers." See Montgomery, C. W. (1997) *Environmental Geology* (5th ed.), The McGraw-Hill Companies, Inc., Boston, p. 219; see also Montgomery, C. W. (2003) *Environmental Geology* (6th ed), The McGraw-Hill Companies, Inc., London, p. 229.

meaning all of those things that are required, necessary or important to (legal) subject(s). Similarly, there are many different definitions for the term "water resource", as revealed below.

In geology, water resource is a "[g]eneral term referring to the occurrence, replenishment, movement, discharge, quantity, quality, and availability of water (non-marine)".[8]

As to academic works as well as policy reports relevant to water resources, only a few of these provide a definition for "water resource".[9] In The New Encyclopaedia Britannica, water resource is defined as "any of the entire range of natural waters that occur on the Earth, regardless of their state (*i.e.*, vapour, liquid, or solid) and that are of potential use to humans".[10] UNESCO/WMO defines water resource as the "[w]ater available, or capable of being made available, for use in sufficient quantity and quality at a location and over a period of time appropriate for an identifiable demand".[11] Under the indicator system of the Food and Agricultural Organization of the UN (FAO), there is a

---

In *The Encyclopaedic Dictionary of Environmental Change*, the term resource is understood as "[a]n anthropocentric concept with qualitative and quantitative meanings: in the qualitative sense, something of use to society, commonly but not necessarily a material; in the quantitative sense, the total amount that exists. Something becomes a resource once demand leads to its exploitation." See Matthews, J. A. ed. (2001) *The Encyclopaedic Dictionary of Environmental Change*, Arnold, London, p. 536.

Soussan, J. believes that, "[a] resource is anything that contributes to the process of production", and [i]n an economic sense, resources do not exist; they are created". See Soussan, J. (1988) *Primary Resources and Energy in the Third World*, Routledge, London, pp. 2-3.

Differently, the *European Community Clearing-House Mechanism* defines resource as "a substance or place required by an organism for its growth, maintenance and reproduction". This definition is available at http://biodiversity-chm.eea.eu.int/CHMIndexTerms/Glossary/R/resource (last visited 31/05/2005).

8 See Bates, R. L. & Jackson, J. A. ed. (1987) *Glossary of Geology*, American Geological Institute, Alexandria, p. 732.

9 For example, the followings do not give a definition to water resource. Marino, M. A. & Simonovic, S. P., ed. (2001) *Integrated Water Resources Management*, Interantional Association of Hydrological Science, Wallingford; Herschy, R. W. & Fairbridge, R. W., ed. (1998) *Encyclopedia of Hydrology and Water Resources*, Kluwer Academic Publishers, London; Pfafflin, J.R., & Ziegler, E. N., ed. (1998) *Encyclopedia of Environmental Science and Engineering* (4th ed.), Gordor and Breach Science Publishers, Australia; Kuiper, E. (1965) *Water Resources Development: Planning, Engineering and Economics*, Butterworths, London; *Agenda 21*; UNESCO, et al. (2003) *The United Nations World Water Development Report: Water for People, Water for Life*, UNESCO Publishing/Berghahn Books, Paris; Organisation for Economic Co-operation and Development (OECD) (1989) *Water Resources Management: Integrated Policies*, Paris.

10 See *The New Encyclopaedia Britannica* (Micropaedia) (15th ed.), Vol. 12, p. 518.

11 See UNESCO/WMO, *supra* note 2.

group of terms regarding the definition of water resource with reference to its quantity.[12] These have been employed broadly across the UN system as well as other organizations and academic institutes.[13] Of these terms, the following are most relevant to this research:

-- "Renewable resources: Total resources that are offered by the average annual natural inflow and runoff that feed each hydrosystem (catchment area or aquifer)."[14]

-- "Total actual renewable water resources (TARWR) (km^3/year): The sum of internal renewable water resources and incoming flow originating outside the country. The computation of TARWR takes into account upstream abstraction and quantity of flows reserved to upstream and downstream countries through formal or informal agreements or treaties. It is a measure of the maximum theoretical amount of water actually available for the country."[15]

-- "Internal renewable water resources (IRWR) (km^3/year): Average annual flow of rivers and recharge of aquifers generated from endogenous precipitation."[16]

-- "External renewable water resources (ERWR) (km^3/year): That part of the country's renewable water resources which is not generated in the country. The ERWR include inflows from upstream countries (groundwater and surface water), and part of the water of border lakes or rivers."[17]

-- "Total natural renewable water resources (km^3/year): The sum of internal renewable water resources and natural incoming flow originating outside the country. It does not vary with time."[18]

2.1.2 Utilisation of Water Resources

The meaning, methods, as well as the process, of the utilisation of water resources, influence the types and substantive content of water rights, and thus it is important to acquire a comprehensive understanding of these terms.

---

[12] See FAO (2003), *Review of World Water Resources by Country*, Rome, pp. xi-xiii.

[13] See the relevant data or statistics employed in *The United Nations World Water Development Report: Water for People, Water for Life* (See UNESCO, et al. (2003) *supra* note 9), the World Bank Group website www.worldbank.org, the World Resources Institute website www.wri.org, and the Pacific Institute website www.pacinst.org, etc.

[14] See FAO (2003), *supra* note 12, p. 11.

[15] *Id.*

[16] *Id.*

[17] *Id.*

[18] *Id.*, p. 12.

2.1.2.1 Water use and its categories

Far as long as human beings have existed, water resources have been utilised for many reasons. However, in different times or in different regions, water resources have been used by different methods and for different purposes, and have played different roles. Even a particular kind of usage may vary over time. The utilisation of water resources for recreation may be taken as an example. As more sources for food have been discovered, fish and wildlife have evolved to be considered more as sources of recreation than sources of food.[19] As a result, the view of recreation from human beings has modified, and it is therefore now considered an important element in the quality of life and has grown into a new industry.[20]

It is true that terms explaining the different uses of water are often employed inconsistently in the water literature, and one should take care when using or reading those terms, *e.g.*, water use, water withdrawal and water consumption.[21] But sometimes the inconsistent use of certain terms is unavoidable, for a particular word may have more than one meaning, or be employed in a broad sense in one place, or in a narrow sense in another, or be viewed from different perspectives. As Gleick, P. H., et al. point out, "[t]he term water use itself encompasses many different ideas", including "[a]mong any other things, it (water use) has been used to mean the withdrawal of water, gross water use, and the consumptive use of water".[22] In the International Glossary of Hydrology, "water use", also referred to as "water exploitation", is defined as the "[u]se of water or altering of its natural condition with the intention of increasing the production of goods and services".[23] In the current work, unless otherwise expressly indicated, the term "utilisation of water resources", "water use" or "using water" are employed interchangeably and in a broad sense, and refer to utilising water to meet certain water needs, consumptively or non-consumptively.

Water uses may be classified according to different criteria. According to the kinds of water being utilised, water uses may be classified as surface water use, groundwater use, atmospheric water use, and conjunctive use, *i.e.*, the "[c]ombined use of surface water and groundwater".[24]

As to the uses of surface water, according to whether or not water is taken from the source during the process of use, water uses can be classified as in-stream use and off-stream use. In-stream use, also referred to as in-channel use,

---

[19] See Beck, R. E., ed. (1991), *supra* note 1, p. 28.
[20] *Id.*, pp. 28-29.
[21] See Gleick, P. H., et al. (1998) *The World's Water 1998-1999*, Island Press, Washington, D. C., p. 12.
[22] *Id.*
[23] See UNESCO/WMO (1992), *supra* note 2.
[24] *Id.*

or non-withdrawal use, refers to "[w]ater use taking place within the stream channel".[25] There are a variety of in-stream uses served today, *e.g.*, water quality improvement, fish breeding, fish and wildlife habitat, recreation, waste assimilation, navigation, hydroelectric power generation, carriage to holders of downstream rights, ecosystem protection, aesthetics, etc., and it is evident that many of these can be provided at the same time.[26] There are also many kinds of off-stream uses, such as irrigation use, domestic use, commercial use, industrial use, electricity production, public supply, and so forth.

Water uses may also be divided into consumptive use and non-consumptive use, according to whether or not the quantity or quality of water being used has been changed after it is used, or whether or not the water is removed from further use. When this change occurs, then there is consumptive use.[27] However, sometimes it is difficult to judge whether a particular use is a consumptive use or a non-consumptive use.

According to the purposes or water users, water uses may be classified into many categories. However, differences exist between different literature, different regions, states and/or countries. For example, in Water and Water Rights (1991 ed.), water uses categories include agriculture, domestic, industry (including steam power generation), commercial and public, fish and wildlife, recreation, hydropower, waste disposal, and navigation.[28] According to the USGS, water uses includes domestic use, commercial use, public-supply use, irrigation use, industrial use, livestock use, mining use, electricity-production use, hydroelectric power use, and wastewater-treatment use.[29] Under the classification used by UNESCO, there are six categories of water use, *i.e.*, agriculture, domestic use, hydropower, industry, inland navigation and

---

25 *Id.*

26 *Id.* See also Beck, R. E., ed. (1991), *supra* note 1, pp. 28-29.

27 In *The International Glossary of Hydrology* (2nd ed.) (See UNESCO/WMO (1992), *supra* note 2), "consumptive use" refers to "[q]uantity of surface and groundwater absorbed by crops and transpired or used directly in the building of plant tissue, together with that evaporated from the cropped area, expressed in units of volume per unit area. Also includes all activities where the use of water results in a loss of the original water supplied, such as industrial or community consumption." Gleick, P. H., et al. think that, "[w]ater consumption or consumptive use should refer to the use of water in a manner that prevents its immediate reuse, such as through evaporation plant transpiration, contamination, or incorporation into a finished product." (See Gleick, P. H., et al. (1998), *supra* note 21, p. 12).

28 See Beck, R. E., ed. (1991), *supra* note 1, pp. 29-37.

29 See USGS, *Water Use in the United States*, available at http://ga.water.usgs.gov/edu/wateruse.html (last visited 31/05/2005).

recreation.[30] Thus, a range of approaches to classifications of different uses exists, which may be confusing.

2.1.2.2 The process of utilisation of water resources

In order to discuss water resources as an object of property law properly, it is helpful to outline the process of the utilisation of water resources, in-stream use as well as off-stream use. As to the process of the utilisation of water resources, in general, there exist differences between in-stream and off-stream use. Figure 2.2 and 2.3 outline the process of the utilisation of water resources, both in-stream and off-stream use. In these two figures, the term "water withdrawal" means the act of taking, drawing or abstracting water from a source of water resources. However, certain steps are not necessarily to be experienced, thus expressed by dashed boxes.

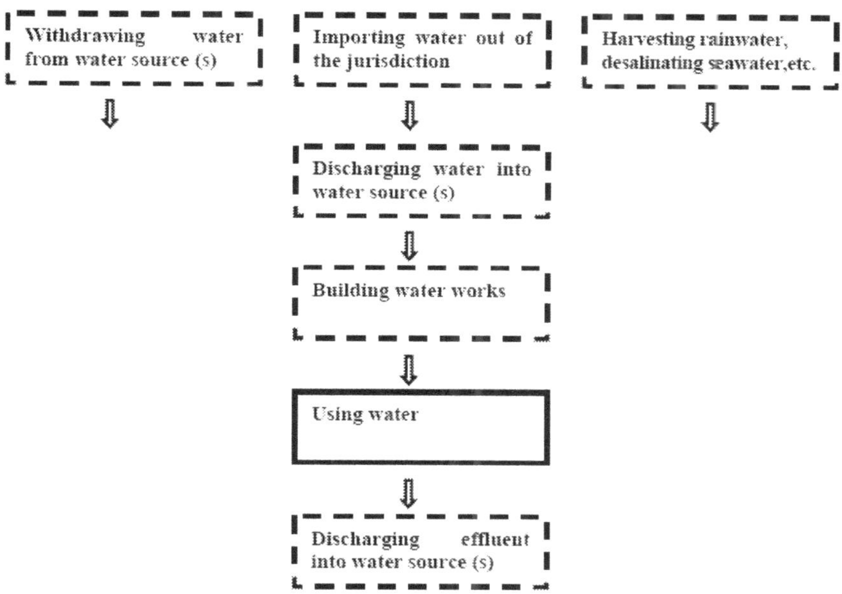

Figure 2.2. Process of Utilisation of Water Resources: In-stream Use. Dashed box expresses a possible step.

30 See UNESCO, *Water use categories*, available at http://www.unesco.org/water/water_links/Water_Issues/Water_use/ (last visited 31/05/2005).

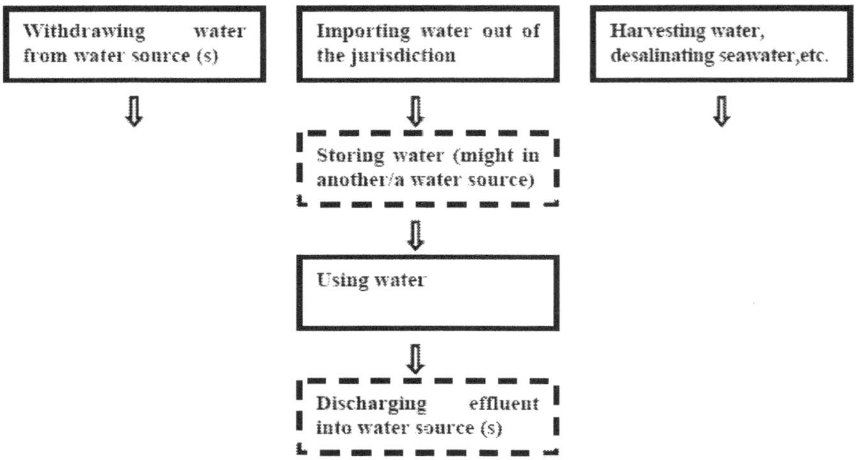

Figure 2.3. Process of Utilisation of Water Resources: Off-stream Use. Dashed box expresses a possible step.

2.1.3 Conclusion

From the above discussion, the author would like to draw the following conclusions:

(1) Freshwater shall be the key object of water resources management for the purpose of this study.

(2) The terms "water" and "water resource" are often used interchangeably to refer to the same thing. However, sometimes it may be necessary to distinguish between these two terms. Although "water resource" commonly refers to the physical freshwater itself, it is also possible that physical water, together with its related surroundings, can form other types of water resources, *e.g.*, hydropower, navigation potential, etc., and this must be considered as part of a water rights scheme. Thus, the properties of water resources, together with level of technology, social and economic development, should determine the utilisation of water resources, including types (category) as well as the process.

(3) All of the above can influence water rights, its structure as well as its legal format and substantial content, and this shall be discussed later in this work.

2.2 EVOLUTION OF WATER RIGHTS

This part will examine the development of water rights through a literature review. One will see that technological, social and economic elements, as well geography and culture, influence the legal forms and substantive content of water rights. The author will analyse the evolution of water rights in two stages - the so-called pre-administration period and the administration period. The beginning of the administration period is the wide intervention or administration of water rights (or the development, allocation and utilisation of water resources) by the executive authorities of the State. However, the beginning of this stage is quite difficult to pinpoint at the global level because of its uneven development worldwide. In contrast, the pre-administration period is characterised by non-intervention or infrequent intervention or administration from the executive authorities of the State. The next part will begin to examine the evolution of water rights.

2.2.1 Water rights: a changing concept

Like many other terms, the word "right" has many different meanings, making it "[a] much ill-used and over-used word".[31] In a general legal sense, the term "right" denotes an advantage, interest, liberty, benefit or privilege conferred or protected by law.[32] More specifically, it has been employed to express a range of legal relations: a privilege to something from the holder of a related obligation, an exemption against infringement of certain fundamental interests, an advantage to do something, a power to initiate a legal relationship, and certain other variations.[33] In proposing a reconciliation of the ideas of economists and lawyers with regard to the term "right", a right has been defined as "[t]he ability to do something which is recognized directly or indirectly in law, to the extent that the law can be expected positively to protect that ability of negatively to prevent others from interfering with that ability".[34] The word "right" used in this work covers claims grounded in the interest of a holder as

---

[31] See Walker, D. M. (1980) *The Oxford Companion To Law*, Clarendon Press, Oxford, p. 1070.

[32] *Id.* See also Vasan, R. S., ed. (1980) *The Canadian Law Dictionary*, Law and Business Publications (Canada) Inc., Don Mills, p. 333; Martin, E. A. (2002) *A Dictionary of Law*, Oxford Press, Oxford, p. 434.

[33] See Hohefeld, W. (1913) Some Fundamental Legal Conceptions as Applied in Judicial Reasoning, *Yale Law Journal* **23**(16), 28-32. See also Waldron, J., ed. (1984) *Introduction to Theories of Rights*, Oxford Press, Oxford; Smart, C. (1989) *Feminism and the Power of Law*, Routledge, London, pp. 138-45; Raz, J. (1984) Legal Rights, *Oxford Journal of Legal Study* 4, 1; Tushnet, C. (1984) An Essay on Rights, *Texas Law Review* 63, 1363.

[34] See Scott, A. and Coustalin, G. (1995) The Evolution of Water Rights, *Natural Resources Journal* **35** (4), 821-979, p. 824.

well as in the public or common interest and the interest of the environment, because sometimes all these interests need to be taken into consideration in an equitable water rights system.

Likewise, the term "water rights" has had different meanings at different stages in history as well as in different regions, states or countries. Writers have defined the term in different ways. For example, Scott, A. and Coustalin, G. think that, "[a] water right can be widely defined as the right to use or enjoy the flowing water in a stream".[35] Beck, R. E., meanwhile, considers that the term water right refers to "a court-protected right to, both now and in the future, use water directly from a natural source, although the right can include the use of water from sources collected through human endeavour as well".[36] "The water right defines how, when and in what amount the water may be withdrawn, as well as what can be done with the withdrawn water, or how it can be used *in situ*."[37] However, each of these approaches reflects the particular developments of political, economic and social life of the country in which the term is defined, meeting the requirements of different culturally and economically developed societies. Some writers point out that, "[w]ith regard to water resources, the policy community consists of law/policy, science/technology and beliefs/ethics. Because of the interrelationships and interdependencies that have evolved over time, changes in any one of these areas will affect the other two".[38] Thus, it is clear that the term "water rights" is an ever-changing concept. In the context of the development of water rights, it is preferred to employ the term "water rights" in its widest meaning, *i.e.*, entitlements relating to water or water related things. Additionally, one should note that, the three aspects of water rights were comprehensively packed with an emphasis on the property right aspect of water resources and there was little attention to the environmental right to water during the pre-administration period, because human rights and, in particular, the right of the environment were not well recognised by the human beings at that time.

---

[35] *Id.*, p. 821.

[36] See Beck, R. E., ed. (1991), *supra* note 1, p. 65.

[37] *Id.*, p. 66.

[38] See Deason, J. P., et al. (2001) Water policy in the United States: a perspective, *Water Policy* (2001) 3, 175-192, p. 176.

2.2.2 Pre-administration system period

2.2.2.1 Water rights under Roman law[39]

Because of the persistent influence of Roman law in the civil law system as well as the common law system, it is first necessary to trace water rights under Roman law. It is well known that the basic classification of Roman civil law was persons, things and actions, and the law of things was concerned with property and obligations. As to water or rivers, they fell into the category of things. According to whether or not could be owned privately, things were divided into two categories, *res in patrimonio* and *res extra patrimonium*. The former were susceptible to ownership in private law and were capable of forming part of the patrimony of a legal person, but the latter could not form part of a private individual's patrimony.

The object matter of the Roman civil law of things was things *in patrimonio*, which were capable of ownership by individuals. However the category of things *extra patrimonium* continued to be of importance.

Res extra patrimonium were subdivided into those *divini iuris*, which were subject to divine law, and *humani iuris*, which were subject to human law. Both of these categories were further divided into three sub-categories.

Res divini iuris consisted of, (a) *res sacrae*, things which were sacred, (b) *res religiosae*, things which were religious, and (c) *res sanctae*, things which were specially protected or sanctioned. *Res sacrae* were things dedicated to the gods above, such as temples, shrines and altars. *Res religioser* were those dedicated to the gods below, the rulers of the world of the spirits of departed ancestors, the *di manes*, and included such things as graves, tombs and funerary monuments. *Res sanctae* consisted of boundary zones, which were placed under special protection because of their peculiar role in dividing what was mine from yours or what was ours from theirs; they included the walls and gates of cities, and other such liminal areas. They were not dedicated to the gods, but were nevertheless under their special protection. Such things could not be owned by a private individual, or a legal person in private law.

As to the other category of things *extra patrimonium humani iuris*, it consisted of:

(1) *res communes*. These were things to all man, *i.e.*, the general property of all mankind. Although such things could not be owned privately, however, the law recognised a right to enjoy them. *Res communes* consisted of the air which one breathed, running water, the sea and the

[39] This part is summarized from Borkowski, A. (1997) *Textbook on Roman Law* (2nd ed.), Oxford University Press, Oxford, pp. 152-156; Watkin, T. G. (1999) *An Historical Introduction to Modern Civil Law*, Dartmouth, Aldershot, pp. 220-221; Scott, A. and Coustalin, G. (1995), *supra* note 34, pp. 833-837.

sea shore between high and low tide. These were common to all mankind and could not be denied to citizens of other states.

(2) *res publicae*. These were things regarded as belonging to a particular state and consisted of perennial rivers, public roads, bridges, harbours, and ports. No citizens could be denied access to their use, but non-citizens could not expect equal access.

(3) *res universitatis*. These were things that were intended for public use, owned by a particular town, city or other community, and were to be enjoyed freely by its inhabitants, such as public streets and buildings, parks, theatres and stadia.

Regarding the rules relating to water resources, they may be summarized as follows:

(1) Running water in streams, which, by its very nature, could not be the object of private ownership, was considered to be *res communes*. It was the general property of all mankind and could not be denied to citizens of other states or be the property of any particular person.

(2) Rivers were classified into "perennial" rivers, *i.e.*, normally flowing all the year round, and "torrential" rivers, *i.e.*, small streams which flowed only in periods of high rainfall or thaw.

(3) All torrential rivers were considered *res in patrimonio*. They were considered to be privately owned by the persons whose land they crossed. If a small stream formed the boundary between two pieces of land, the owners of that land owned it to its midway mark. The right to divert water in these rivers was private, belonging to the owner of the land. The state or public had no claim to the banks of these private rivers.

(4) All perennial rivers were considered to be *res publicae*. They were the general property of all the citizens of a particular state. No citizens could be denied access to their use; but non-citizens could not expect equal access. The state owned the bed of the river as well as its banks. These rivers were subject to the authority of a centralized administration, and public rights of navigation, fishing and access were emphasized.

(5) The right to divert water in perennial rivers was available to the public, subject to certain state regulations.

(6) A person could obtain a temporary usufructurary right to the diverted water, and the right lasted only as long as the specific use continued. The right to divert water was irrespective of an individual's ownership of land. However, to exercise such a right needed a prerequisite, *i.e.*, he would have to have or acquire legal access to the banks.

(7) A "prescriptive" right to divert a certain quantity of water in perennial rivers from one year to the next could be acquired by prolonged and unchallenged action over a period of years. After this time, the state could no longer challenge the right as long as it remained the same.
(8) It was permissible to transport water out of the stream land. The law recognized easements (servitudes) whereby a person granted another the right to lay pipes across his land to access the river and draw out water.
(9) Law concerning damage stated that one must not hinder another's use of water, nor harm his property. Therefore, an existing use had priority over a later use.

2.2.2.2 The riparian rights doctrine

Introduction

The fundamental concept of the riparian rights doctrine is that the proprietor of the land abutting a surface water body is entitled to the enjoyment of a bundle of legal interests to the water.[40] In other words, water rights under the riparian rights doctrine are a series of land-based rights.

The origins of the riparian rights doctrine are debatable. There are three different viewpoints. The first holds that the doctrine has developed from Roman law through France. The second asserts that the riparian rights doctrine was rooted in English common law. And the third insists that it originated in the courts of the United States of America (USA).[41] Despite these different theories, the following two facts should not be ignored: (a) Roman law has influenced common law, including English common law, US common law and Canadian common law;[42] and (b) some aspects of water law in early France are similar to that of riparian rights doctrine under common law. The common law system originated in the United Kingdom of Great Britain and Northern Ireland (UK), but one must also take note of its development in the USA. When we trace water rights back under common law, it is necessary and of great value to study the relevant history in two areas, *i.e.*, the UK and the USA.

Prior to the 18th century, most of the water cases in common law countries involved rights of navigation and fishing.[43] The Industrial Revolution in Europe and North America, especially in the USA, led to an increase in power demand, and waterpower was included as a competing right of use. Over 500,000 mills existed throughout Europe by 1800, many of which included a mill raceway or

---

[40] See Beck, R. E., ed. (1991), *supra* note 1, pp. 87-89; Getches, D. H. (1997) *Water Law in a Nutshell* (3rd ed.), West Publishing Co., St Paul, p. 15; Howarth, W. (1992) *Wisdom's Law of Water Courses* (5th ed.), Shaw & Sons Limited, Crayford, p. 66.

[41] See Beck, R. E., ed. (1991), *supra* note 1, p. 233; Getches, D. H. (1997), *supra* note 40, p. 15.

[42] See Scott, A. and Coustalin, G. (1995), *supra* note 34, pp. 833-835.

[43] See Getches, D. H. (1997), *supra* note 40, p. 16.

canal, that diverted water from the river to provide an efficient and steady flow of water for a waterwheel; mills powered by waterwheels ground corn and wheat, powered bellows and hammers to make iron, ground ingredients to make paper, cut wood, crushed olives for oil, drilled gun barrels, and powered textile factories.[44] This situation led to more conflict of use between different kinds of use as well as different users of the same use, and led to a need for a uniform set of legal principles that could settle the growing number of water disputes about access to the flow of the stream, from which arose the riparian rights doctrine.[45]

Under common law, two different systems of water law, *i.e.*, the riparian rights doctrine and the prior appropriation doctrine, developed before 1900 to allocate uses of shared waters. Following the increasing multiple utilisation of water resources, and the trend of enactment of statutes in law-making, the two systems have gradually evolved and existed not only in statutory form, but also have combined to some extent.

Early French water law

With the Industrial Revolution, the owners of French industry wished to control the state power, which finally led to the Great Revolution. In response to the need to unify the law across the entire country, the Civil Code was promulgated in 1804. Under the Civil Code 1804, there were three kinds of categories relating to water, and different rules were applied.

The first classification under the French Civil Code 1804 covered rivers or streams of the public domain. In the Code, "rivers or streams which will carry floats ... which are not susceptible of private proprietorship are considered as dependencies on public domain".[46] They were subject to administration and should be used in accordance with the forms and pursuance of regulations applied to them.[47]

The second category covered running water other than rivers or streams in the public domain. As to this kind of water, two different circumstances existed. The first was where a property bordered such water, and the proprietor could water his property with this resource. The second was when an owner's land was intersected by this kind of water; in such a case, the owner was entitled to use water at his pleasure "within the space through which it runs, but on condition of restoring it, at the boundaries of his field, to its ordinary course".[48]

---

44 See Cech, T. V. (2003) *Principles of water resources*, John Wiley & Sons Inc., New York, p. 18.

45 *Id.*

46 See art. 538, French Civil Code 1804. French Civil Code 1804 (English version) available at http://www.napoleonseries.org/reference/political/code.cfm (last visited 31/05/2005).

47 *Id.*, art. 537.

48 *Id.*, art. 644.

The third category related to springs. When a spring was located in a field, the possessor could make use of it at liberty, except (a) when the owner of a lower field had acquired the right to use the water by title or by prescription, in which case, such a right had be protected; and (b) when a spring supplied water to the residents of a commune, village or hamlet for their necessary use, the owner of the spring could not change the course of the spring, although he could claim an indemnity whenever the residents had not acquired the right to use the water.[49]

As to the right to fish, this was treated as one mode of acquiring property, and was regulated by particular laws.[50]

On 8th April 1898, a new law was enacted with amendments and/or supplements to the above provisions. There were four relevant articles as follows:

> "Article 641(1) Any owner has the right to use and dispose of rainwaters which fall on his land.
>
> Article 641(3) The same provision is applicable to waters of springs originating on a land.
>
> Article 642(1) One who has a spring on his land may always use the waters of it as he wishes within the limits and for the needs of heritage.
>
> Article 643 If, after leaving the land where they arise, waters of a spring form a watercourse presenting the character of public and running waters, the owners may not divert them from their natural course to the prejudice of lower users." [51]

It should be noted that, under the latter two circumstances, the right of using water of proprietors in running water other than rivers or streams was on public domain, and the lower proprietor and the residents in springs belonged to a kind of servitudes derived from the situation of places, and was adjusted by general provision of servitudes derived from the situation of places provided by article 641.[52] On the other hand, groundwater was not stipulated clearly by the Code, while article 552 provided that "[p]roperty in the soil imports property above and beneath".[53]

Early riparian rights doctrine – the natural flow theory

---

49 *Id.*, arts. 641-643.

50 *Id.*, art. 715.

51 See Crabb, J. H. (1995) *The French Civil Code - Revised Edition (as amended to 1 July 1994)*, Fred B. Rothman & Co., Littleton.

52 See art. 640, French Civil Code 1804. It provides that, "Inferior lands are subjected, as regards those which lie higher, to receive the waters which flow naturally therefrom to which the hand of man has not contributed. " "The proprietor of the lower ground cannot raise a bank which shall prevent such flowing." "The superior proprietor of the higher lands cannot do any thing to increase the servitude of the lower."

53 *Id.*, art. 552.

Before and during the early stages of the Industrial Revolution, agriculture was at the heart of social life in England as well as in New England and later eastern USA, and the importance of land is therefore emphasised. Water uses, *e.g.*, navigation, a small quantity of waste disposal, turning waterwheels for generating power in or near streams, irrigation without fertilizers or pesticides, small scale mining, etc., were all considered to be non-consumptive uses because these had little or no effect on the quantity and quality of the water in rivers except for a small quantity of domestic use, on the one hand. On the other hand, in both England and eastern USA, rainfall was plentiful enough to meet the needs of all types of water use, even commercial agriculture, as it then existed. The relatively abundant surface water and the lack of technology to draw groundwater made the utilisation of surface water closely tied to the land, and riparian rights were involved. As to the use of water incidental to agriculture, there existed limited manorial rights to use the river's attributes for ferries, navigation, fishing and waterpower. The few water mills there were operated mainly to saw wood, grind corn, full yarn, or pound ore and metal. However, the abundant water resources compared to the limited, simple but unpolluted water use, did not give rise to many disputes between river users, and as a result, simple rules were adequate to allocate the use of water resources. During this period, the natural flow theory originated, and was developed under the riparian rights doctrine. According to the natural flow theory, "[e]ach riparian owner on a (surface) waterbody is entitled to have the water flow across, or lie upon, the land in its natural condition, without alteration by others of the rate of flow or the quantity or quality of the water".[54]

Later riparian rights doctrine – the reasonable use theory

With the development of the Industrial Revolution, especially with the invention of the steam engine, which greatly promoted the Industrial Revolution, the demand for water and waterpower increased hugely. The new industry needed more waterpower to turn the mills for falling corn, weaving and spring textiles and grinding ores, and the mills became larger and larger, spread further apart and were sometimes located far away from rivers.[55] However, the processes returned to the river not just water, but also heat, chemical and organic waste, and the river pollution became a new problem to be dealt with. More and more water uses altered the natural state of the river.[56] As a result, disputes increased among riparians, and the natural flow theory constituted a

---

[54] See Beck, R. E., ed. (1991), *supra* note 1, p. 233. See also Getches, D. H. (1997), *supra* note 40, p. 4.

[55] See Getzler, J. (2004) *A History of Water Rights at Common Law*, Oxford University Press, Oxford, pp. 23-39.

[56] *Id.*

threat to the commercially valuable development of water resources.[57] That is to say, the natural flow of the stream began to give way to the need to alter the stream to maximize water use, and this led to a demand for rivers to be developed to maximize their economic value, as well as to let them flow in a reasonable state.[58] In this situation, the reasonable use theory developed to displace the natural flow theory.

The famous decision in the case of *Tyler v. Wilkinson* in 1827 led to the development of the reasonable use theory.[59] In this decision, Justice Story insisted that in fact neither party was entitled to the natural flow of the river, because any use of the river causes some grade of retardation, acceleration or diminution of the flow, and so the injury of others. However, a proper balance should be found, that is the extent of the injury that can be tolerated. A riparian must not be forbidden from making a valuable good or an enjoyment of the flow merely for the reason that another riparian might incur increasing cost or inconvenience. In other words, while all riparians were equally entitled to use the water in a naturally flowing stream, each one had the right to make a reasonable use of the water without interfering with the reasonable use of any other riparian. Naturally, this decision changed the preceding riparian rights doctrine that forbade the riparian to modify the natural flow of the river. It is clear that this ethic "explicitly protected industrial exploitation of the rivers to a certain 'reasonable' extent".[60] In the following year, Chancellor Kent discussed the law relating to water rights in the third edition of his Commentaries on American Law, in which the reasonable use doctrine was upheld and Story's decision was distinctively quoted along with lots of civil law authorities.[61] After that, gradually, the court in the USA and in England followed Kent's Commentaries and Story's verdict in their deliberations.[62]

Under the reasonable use theory, "each owner of riparian land is permitted to use the water in a (surface) waterbody, regardless of the effect the use has on the natural flow, so long as each user does not transgress the equal right of other riparians to use the water".[63] In other words, the later development of riparian rights doctrine is subject to reasonable use. The principal riparian rights are:

(1) The right of access to the river;[64]

---

[57] *Id.*, pp. 39-41.

[58] *Id.*, p. 44.

[59] 24 F. Cas. 472 (D.R.I. 1827) (No. 14,312).

[60] See Scott, A. and Coustalin, G. (1995), *supra* note 34, p. 873.

[61] See Getzler, J. (2004), *supra* note 55, p. 281. See also Getches, D. H. (1997), *supra* note 40, p. 20.

[62] *Id.*

[63] See Beck, R. E., ed. (1991), *supra* note 1, p. 241. See also Getches, D. H. (1997), *supra* note 40, p. 4.

[64] See Macrory, R. (1985) *Water Law: Principles and Practice*, Longman Professional London, p. 9; Getches, D. H. (1997), *supra* note 40, p. 33.

(2) The right to the flow of the river;[65]
(3) The right to make a reasonable use of the waterbody, *e.g.*, the right to abstract or divert water, the right to impound or obstruct water;[66]
(4) "The right to have the water flow in a 'natural' state of quantity";[67]
(5) The rights of navigation, fishing, etc.[68]

These are land-based rights, and have the following five features:

> "-- only the owner of the stream banks has the rights in the flow;
>
> -- rights are to the undiminished, unaltered flow of the water;
>
> -- the owner may use the flow at any time, in any way, and in any quantity, provided he does not cause damage to other owners along the stream (that is, alter or diminish the flow to them);
>
> -- rights are relative and impose corresponding obligations amongst the riparian community. The rights and obligations of each riparian are equal, regardless of how much land he owns;
>
> -- full riparian rights can be transferred only by transferring title to the riparian land."[69]

2.2.2.3 The prior appropriation doctrine

The situation in the vast western USA is quite different from that in the east. Firstly, the west belongs to the arid region beyond the one-hundredth meridian. Secondly, the federal government owned the majority of land in the western USA. Thirdly, the west coast is rich in timber, furs and minerals, and suitable for farming. The policy of the federal government to encourage settlement in the west led to development in that region. However, the most promising mineral deposits were often far from any stream. The federal government acquiesced in the settlers' trespasses and use of water. Therefore, the settlers developed their own customary rule, which was different from the riparian rights doctrine, *i.e.*, (a) quantities of water could be diverted from a stream, transported many miles and put to use for mining or irrigation; (b) this right to appropriate a quantity of water was granted on a "first come- first served" basis; and (c) anyone could make a diversion that did not deprive "prior appropriators" of the quantity of water already being diverted by them. Gradually, the prior appropriation

---

65 *Id.*
66 *Id.*
67 *Id.*
68 *Id.*
69 See Scott, A. and Coustalin, G. (1995), *supra* note 34, p. 825.

doctrine was adopted by state courts and finally has been incorporated into state statutes.[70]

Compared to the riparian rights doctrine, water rights under the prior appropriation doctrine are use-based rights, and the basic premises are as follows:

> "-- rights exist only as long as water use exists. Land ownership is not essential provided there is legal access to the river;
>
> -- rights are specific as to quantity and type of use;
>
> -- the first user has the strongest rights. A seniority system provides ranking of rights;
>
> -- rights-holders (users) can enforce their rights only against those lower in seniority (later in time);
>
> -- the usufructuary rights are fully transferable to any person."[71]

2.2.2.4 Conclusion

During the pre-administration system period, water law concentrated on the adjustment of surface water,[72] and emphasised the right of access to water, with the ownership of physical water being of lesser importance. Notably, different circumstances, including social and economic development, as well as climate, cultural and geological elements, resulted in the development of very different water law regimes. With the development of types of water use as well as increasing demand in terms of both quantity and quality, plus the increasing impact of every kind of water use on water quality as well as on hydro-ecology, the drawbacks of the water law regimes during the pre-administration period gradually began to emerge.

The riparian rights doctrine could "be exceedingly vague".[73] In particular, its defects include the following four aspects: (a) the inability to quantify water rights. This could lead to the rights being under uncertainty in terms of quality, and therefore this is unhelpful in encouraging investment and led to increasing difficulties in transferring or modifying rights;[74] (b) the limitation on water being transferred to higher valued uses;[75] (c) the ineffectiveness in protecting the environment;[76] (d) the invalid assumption. The assumption of the doctrine is

[70] See Getches, D. H. (1997), *supra* note 40, pp. 20-22.
[71] See Scott, A. and Coustalin, G. (1995), *supra* note 34, p. 825.
[72] See Macrory, R. (1985), *supra* note 64, pp. 5-6.
[73] See Deason, J. P., et al., *supra* note 38, p. 177.
[74] See Scott, A. and Coustalin, G., *supra* note 34, p. 960; see also Deason, J. P., et al., .*supra* note 38, p. 177.
[75] See Deason, J. P., et al., *supra* note 38, p. 177.
[76] *Id.*

the existence of an abundant water supply. However, circumstances have changed greatly.[77]

Comparatively, the prior appropriation doctrine could "be exceedingly rigid".[78] The following three aspects should be noted in this regard: (a) the unbending of the use of the water required under the seniorities. This prevents water from being transferred to higher valued uses;[79] (b) the insufficiency to protect the environment. Water use that is difficult to enumerate economic profits, *e.g.*, minimum flow and other in-stream uses, could not be protected effectively;[80] (c) general "indifference to river attributes other than the flow and its user group which floats without specified location or long-term direction".[81]

In addition to the above, there is also a vital shortcoming under the riparian rights doctrine as well as the prior appropriation doctrine in the changed situation, which relates to distinguishing *the right to use water* from *the right of ownership in that water*. As to their difference, Parke, B., in *Embrey v. Owen* made the following, often cited, statement:

> "the right to have a stream flow in its natural state, without diminution or alteration is an incident of property in the land through which it passes; but flowing water is *publici juris*, not in the sense that it is *bonum vacans*, to which the first occupant may acquire an exclusive right, but that it is public and common in the sense only, that all may reasonably use it who have a right of access to it , and that none can have any property in the water itself, except in the particular portion which he may choose to abstract from the stream and take into his possession, and that during the time of his possession only."[82]

That is to say, *the right to use water* is more important than *the right of ownership in that water*.[83] This distinction might be proper in circumstances where little water is being abstracted from a natural source and taken into possession, and non-consumptive water use is dominant. However, with the increasing amount of water being abstracted and used consumptively, the problems relating to the vague meaning of the word "use" appeared. In conclusion, the changed circumstances required the removal of these drawbacks, and therefore it is necessary to modify or revise the existing water law systems.

---

77 *Id.*

78 *Id.*

79 *Id.*

80 *Id.*

81 See Scott, A. and Coustalin, G., *supra* note 34, p. 960; see also Deason, J. P., et al., .*supra* note 38, p. 177.

82 See Per Parke, B., in *Embrey v. Owen*, (1851) 6 Ex. 353.

83 See, Howarth, W. (1992), *supra* note 40, p 67.

2.2.3 Administration system period

Society is constantly evolving, and industrialization in all regimes has promoted great economic development. At the same time, water demands have increased, water pollution has led to serious threats, the scarcity of water resources has become more and more severe, and conflicts over water use have increased. In this content, neither the riparian rights doctrine nor the prior appropriation doctrine could resolve or prevent many water-related problems, because of their particular shortcomings, as noted above. Gradually, an administrative system with a water drawing (abstraction) system at its core has developed. For example, in the USA, both the riparian rights doctrine and the prior appropriation doctrine have evolved into the modern statutory system.[84]

In the UK, the riparian rights doctrine has become subject to, and been extensively modified by, a range of statutory provisions dealing with abstraction, pollution control and various other matters.[85] The adoption of the Water Resources Act 1963 marked the establishment of an administrative system of water rights in England and Wales.[86] Under this administrative system, anyone intending to take water from or use water in a natural source is required firstly to obtain a permit or licence from the relevant department of government, despite certain exemptions.[87]

In the USA, the administrative statutes relating to water rights have been influenced by the riparian rights doctrine, as well as the prior appropriation doctrine, and, to some extent, the two doctrines have become entwined.[88] For example, some of the riparian rights, *e.g.*, the right of purity, to fish, to have access to the stream, and to protect the banks of a steam from erosion, have been incorporated into the permit system of the western USA.[89] As Robert E. Beck points out, "[m]ore and more, both in the eastern and western USA, the water right is reflected by a piece of paper; a license or a permit".[90]

Up to this point, the author would define the main characteristics of water law in the administrative system as:

(1) Water resource including groundwater as well as surface water, generally in its natural state, is to be considered as part of the public domain and is subject to the States' administration;

(2) The holding of a permit or licence is a precondition of access to water, while there are certain exceptions and/or exemptions;

---

84 See Getches, D. H. (1997), *supra* note 40, p. 22, 27.

85 See Howarth, W. (1992), *supra* note 40, pp. 66-67.

86 See the long title of the Water Resources Act 1963 (c. 38).

87 *Id.*, sections 23 and 24.

88 See Deason, J. P., et al.(2001), *supra* note 38, pp. 177-178.

89 See Getches, D. H. (1997), *supra* note 40, p. 22.

90 See Beck, R. E. ed. (1991), *supra* note 1, p. 66.

(3) Water use should be in accordance with the conditions laid down by the law, permit or licence;
(4) In some countries, or some states of some federal countries, a water permit or a licence may be transferred, in accordance with the law;
(5) The permit or licence system is usually an important method or component of water administration;
(6) Generally, the private ownership in physical water exists only after water has been abstracted by private individuals in accordance with the permit or licence. In some jurisdictions (*e.g.*, South Australia[91]), this is recognized expressly by law, while in others (*e.g.*, China[92]), private ownership of physical water exists in fact.

2.3 WATER RIGHTS: THE PROPERTY RIGHT ON WATER RESOURCES

In order to provide an overall concept of the "property right" on water resources, this part will firstly analyse the objects of property rights, including water resources. The work will then discuss the legal forms and content of the property right on water resources, as a key element of the water rights discussion that lies at the heart of this work.

"At the most basic, the term property is simply a more technical expression describing 'things'."[93] However, in the legal sense, the term "property" can be invoked to express two relevant but distinct meanings, *i.e.*, a) the right of ownership, and, more commonly, b) the object of the right of ownership.[94] It may be stated that the property law, or the law of property, under the civil law system is more logical in structure than that under the common law system,[95] which is why many legal intellectuals in common law jurisdictions have looked to for "received, reinvented, applied, discarded, simplified, and adapted" ideas from Roman law without "entirely mindless, unselfconscious, or mechanical jurisprudence".[96] However, the common law property system may still yield valuable information from which China may benefit, *e.g.*, its rules on trust.

---

91 See the relevant discussion in 5.3.3 of Chapter 5.

92 See the relevant discussion in 3.4.1 of Chapter 3.

93 See Pearce, R. A. & Stevens, J. (1998) *The Law of Trusts and Equitable Obligations* (2nd ed), Butterworths, London, p. 86.

94 See Walker, D. M. (1980), *supra* note 31, p. 1007. See also Ganer, B. A., ed. (1999) *Black's Law Dictionary* (7th ed.), West Publishing, St. Paul, p. 1232.

95 See Atiyah, P. S. (1987) *Pragmatism and Theory in English Law*, Stevens & Sons, London, pp. 9-10.

96 See Getzler, J. (2004), *supra* note 55, pp. 328-331.

In accordance with the *numerus clausus* doctrine of property law under the civil law system, legal rights and obligations pertinent to property are and should be expressly stipulated by law.[97] Therefore, the object of the right of ownership must be things, but not all things are the object of the right of ownership.[98] The term "property" will be employed here to refer to "things". In order to make the legal provisions logical, concise, or of a certain abstract, the law and theory of law usually differentiate between different types or categories of properties. The property may be classified in different ways, *e.g.*, tangible and intangible, corporeal and incorporeal, movable and immovable, etc.[99] There are two kinds of physical or tangible things in the world, (a) natural things, *i.e.*, things that originate from the nature, and (b) artificial things, *i.e.*, things that derive from human activities. In general, when a kind of natural things is plentiful enough, difficult for humans to possess, or is of limited economic value, it is unnecessary for property law to regulate them; in other words, it lies beyond the scope of objects of property law. The air may be taken as an example.[100] In contrast, there exist property rights over artificial things in principle. The property right is a group of rights, including the right to occupy, possess, use, abuse, lend, let out, alienate, use up, consume, transfer in security, sell, exchange, gift, bequeath, destroy, etc., and two or more of these rights sometimes combine to form other kinds of property rights in law.[101] For example, when all of these rights combine together, ownership will be formed, and when some combine, they form a usufructary or servitude. The "ownership is the most absolute power permitted by law to enjoy or dispose of things of the kind in question", and some or even many rights may be relinquished temporarily without loss of ownership.[102]

As to water resources, because of its character as a social good, in most countries, water resources in its natural state lies under public domain.[103] In

---

97 See Sjef van Erp (2003) A Numerus Quasi-Clausus of Property Rights as a Constitutive Element of a Future European Property Law? *ELECTRONIC JOURNAL OF COMPARATIVE LAW* Vol. 7.2 June 2003, available at http://www.ejcl.org/72/art72-2.html (last visited 31/05/2005).

98 See Walker, D. M. (1980), *supra* note 31, p. 1007.

99 *Id.*, p. 1007, 1216; Ganer, B. A., ed. (1999), *supra* note 94, pp. 1233-1234, 1488. See also, the explanation of "THINGS" in *Bouviers Law Dictionary* (1856 ed.), available at http://www.jusbelli.com/Bouvier/bouvier1856_th.html (last visited 31/05/2005).

100 See Walker, D. M. (1980), *supra* note 31, p. 1007.

101 *Id.*, p. 1008.

102 *Id.*

103 See Solanes, M. & Gonzalez-Villarreal, F. (1999) *The Dublin Principles for Water as Reflected in a Comparative Assessment of Institutional and Legal Arrangements for Integrated Water Resources Management*, Global Water Partnership, Stockholm, p. 29.

some countries, only the State has property ownership over water resources, and private individuals, thus, can only use or get water resources in accordance with the law.[104] In other countries, such as China, the State has ownership over surface water as well as groundwater, regardless of whether or not the water is in its natural state.[105] Under Chinese water law, individuals only have the right to use water according to strict provisions of the law.[106]

Under either the riparian rights doctrine or the prior appropriation doctrine, water rights may virtually be treated as the right to use water (mainly for in-stream use) or the right to take water (mainly for off-stream use).[107] In the first case, it is the right of usufructary, and in the second, it is the right to obtain property rights over physical water, although it has a certain direct economic value in some cases. However, the author would assert that the private property right over physical water should be emphasized, and a permit or licence should be considered as a mode or a right for acquiring property rights from the viewpoint of property law, and the permit or licence system should be considered as a method for administering and/or allocating water resources in its natural state, although a permit or licence may be of economic value sometimes. In brief, the content of the property right on water resources consists of the ownership on water resources, and the right to use water as well as the right to withdraw water.

2.4 WATER RIGHTS: THE HUMAN RIGHT TO WATER

2.4.1 Introduction

By analysing the vital role of water for human beings, this part will examine why and how the right to water is a human right.

"Chemically, the human body consists mainly of water and of organic compounds", and it "is about 60 percent water by weight".[108] In other words, the human body is mostly water. Moreover, "[w]ater is the essence of life. Without water, human beings cannot live for more than a few days. It plays a vital role in nearly every function of the body, protecting the immune system –

---

104 *Id.* Regarding examples from legislation from different jurisdictions, see the relevant discussion in 6.1 of Chapter 6.

105 See art. 3, China Water Law 2002. It provides that, "[w]ater resources shall be owned by the State. The ownership of water resources shall be exercised by the State Council on behalf of the State".

106 *Id.*, art. 6.

107 See the previous discussion in 2.2.2 of this Chapter.

108 See *The New Encyclopaedia Britannica (Micropaedia)* (15th ed.), Vol. 6, p. 134.

the body's natural defences – and helping remove waste matter".[109] It is clear that people could not survive without water. In this sense, the right to water for personal and domestic use shall be viewed as a right under natural law.[110] The writer believes that access to water to meet basic human needs is a human right, *i.e.*, a right "inherent in individuals as rational, free-willing creatures, not conferred by mere positive law, nor capable of being abridged or abrogated by positive law".[111] As the Committee on Economic, Social and Cultural Rights (CESCR) of the UN Social and Economic Council has declared, "[t]he human right to water entitles everyone to sufficient, safe, acceptable, physically accessible and affordable water for personal and domestic uses".[112]

The human right to water is a natural right common to all human beings, with the exception of where the right is abandoned voluntarily. However, whenever or wherever there is plenty of water for human beings to enjoy easily, it is not necessary for the law to regulate the human right to water specially, as may be easily understood from the evolution of water rights, as discussed above. With the increasing demand for water, together with the serious water pollution, water has become scare in many areas of the world, and this situation has led to many serious problems. According to the WHO, 1.1 billon persons lack access to

---

[109] See World Health Organization (WHO) (2003) *The Right To Water*, WHO, p. 6.

[110] Under the classical natural law theory, Aquinas points out, "all humanly enacted laws are in accord with reason to the extent that they derive from the natural law. And if a human law is at variance in any particular with the natural law, it is no longer legal, but rather a corruption of law"; see D'ENTRÈVES, A. P., *AQUINAS SELECTED POLITICAL WRITINGS*, translated by DAWSON, J. G., Basil Blackwell, Oxford, 1965, p. 129. Blackstone, W. describes the thesis, "This law of nature, being co-eval with mankind and dictated by God himself, is of course superior in obligation to any other. It is binding over all the globe, in all countries, and at all times: no human laws are of any validity, if contrary to this; and such of them as are valid derive all their force, and all their authority, mediately or immediately, from this original"; see Blackstone, W. *Commentaries on the Law of England (1765-1769)*, the University of Chicago Press, 1979, p. 41.

[111] See Walker, D. M. (1980), *supra* note 31, p. 591.

[112] See para. 2, General Comment No. 15 (2002): The right to water (arts. 11 and 12 of the International Covenant on Economic, Social and Cultural Rights) (hereinafter "General Comment 15"), 26 November 2002, UN Doc. E/C.12/2002/11; also available at UN website http://193.194.138.190/html/menu2/6/gc15.doc (last visited 31/05/2005). Also, in para. 3, the General Comment 15 points that, "Article 11, paragraph 1, of the Covenant specifies a number of rights emanating from, and indispensable for, the realization of the right to an adequate standard of living 'including adequate food, clothing and housing'. The use of the word 'including' indicates that this catalogue of rights was not intended to be exhaustive. The right to water clearly falls within the category of guarantees essential for securing an adequate standard of living, particularly since it is one of the most fundamental conditions for survival".

basic water supply, and 2.4 billion persons are without proper sanitation.[113] In this circumstance, the right of access to water to meet basic human needs has been discussed as a human right since the 1970's, especially since the 1990's. To date, the human right to water has been recognized, embodied or provided explicitly or implicitly in a wide range of relevant international documents as well as in domestic documents. This part will discuss the State practice concerning the human right to water, and delay the discussion of its recognition in international law until Chapter 4.

2.4.2 State practice

When discussing the human right to water, few people consider that there exist sufficient state practice to support this right, and so this question is seldom discussed.[114] The author believes that the state practice relating to the human right to water should be discussed from two perspectives, *i.e.*, law and policy, although sometimes a particular country employs both of them. Further, one should not only review whether the term the human right to water is employed, but also analyse the content of the domestic law or policy of a particular country to conclude whether the human right to water is regulated in that country. As will be revealed, the relevant state practice is strong enough to support the human right to water, although in some countries the human right to water is explicitly stipulated while in others it is implicitly provided.

2.4.2.1 Domestic law

At the national level, certain countries provide the human right to water explicitly. In South Africa for example, the human right to water is clearly stipulated in the constitution as well as in water law, and this provides a good example.[115] The Constitution of Zambia 1996 provides that, "[t]he State shall endeavour to provide clean and safe water … for all persons, and take measures to constantly improve such facilities and amenities".[116] The Constitution of

---

113 See WHO (2000), *The Global Water Supply and Sanitation Assessment 2000*, WHO, Geneva, p. 1.

114 This could be illustrated by some leading or famous documents or works such as follows: *General Comment No. 15*; WHO (2003), *supra* note 109; Salman, M. A. S. and Mclnerney-Lankford, S. (2004) *The Human Right to Water: Legal and Policy Dimensions*, World Bank Publication, Washington, D.C.; Scanlon, J., et. al. (2004) *Water as a Human Right?* World Conservation Union.

115 See the detailed discussion in 5.2.3 of Chapter 5.

116 See art. 112 (d), Constitution of Zambia 1996, its English version available at http://www.electcom.org.zm/constitutionofzambia.html (last visited 31/05/2005).

Ethiopia 1998 declares that, "[t]o the extent the country's resources permit, policies shall aim to provide all Ethiopians access to ... clean water...".[117]

Other countries regulate the human right to water through implicit methods, *e.g.*, giving priority to household water use, taking water for household use without a permit or licence, etc. In China, water law gives first priority to the water demands of urban and rural inhabitants, and the act of drawing a small amount of water for household use is among the exemptions from water-drawing licence system and water resources paid use system.[118] The French Water Act 1992, meanwhile, lists among its objects the ensuring of "the supply in drinking water to the population" as one object.[119] Under the UK Water Resources Act 1991, the restriction on abstraction shall not apply to the domestic purposes in general conditions.[120] According to the Germany Act on Managing Water Resources 2000, for domestic purpose water uses, "[a] permit or an approval shall not be required for the removal, bringing to the surface, guiding to the surface or diverting of groundwater".[121] In South Australia water law, a person who has lawful access to water resources is entitled to take water for domestic purposes from the relevant sources without a licence, even where the relevant sources is prescribed.[122]

2.4.2.2 Domestic policy

Regarding domestic policy, Agenda 21 has exerted a major influence. In accordance with Agenda 21, "[i]n developing and using water resources, priority has to be given to the satisfaction of basic needs and the safeguarding of ecosystems".[123] One should note that almost all States have adopted the relevant agenda 21 at national level, and the relevant local agenda 21 has been enacted at

[117] See art. 90(1), Constitution of Ethiopia 1998, its English version available at http://www.ethiopar.net/English/cnstiotn/consttn.htm (last visited 31/05/2005).

[118] See the detailed discussion in 3.5.1 of Chapter 3.

[119] See art. 2, French Water Act of January 3, 1992, Law No. 92-3 of January 3, 1992, On Water, ("Journal official" Jan. 4, 1992, p. 187). Its English version translated by the French Ministry for the Environment, and available at http://www.oieau.fr/anglais/gest_eau/lois/watlaw92.htm (last visited 31/05/2005).

[120] See sections 27 (4) (a) (i), and (5), Water Resources Act 1991 (c. 57), available at http://www.legislation.hmso.gov.uk/acts/acts1991/Ukpga_19910057_en_1.htm#tcon (last visited 31/05/2005).

[121] See art. 33 (1), Germany Act on Managing Water Resources 2000, Federal Law Gazette (2000) I, p. 632; its English version available at http://www.iuscomp.org/gla/statutes/WHG.htm (last visited 31/05/2005).

[122] See the relevant discussion in 5.3 of Chapter 5.

[123] See para. 18.8, *Agenda 21*.

various local levels in many States.[124] Therefore, it may be concluded that the human right to water has been recognised in domestic policies.

Secondly, the human right to water may be found in the domestic water policies of many countries. South Africa continues an excellent case, and will be discussed in detail in Chapter 5.[125] In India, the National Water Policy 2002 definitely points out, "[w]ater is a prime natural resource, a basic human need and a precious national asset",[126] and emphasises that the drinking water needs of people must be met.[127] In the Mozambique National Water Policy 1995, the first of the main policies is the basic human need: "The satisfaction of basic needs is a high priority and will require an increased coverage of water supply and sanitation, in particular to rural, low income groups".[128]

In the Bangladeshi National Water Policy 1999, the priority of domestic water use is declared. "In general, the priority for allocating water during critical periods in the water shortage zones will be in the following order: domestic and municipal uses, non-consumptive uses (e.g. navigation, fisheries and wild-life), sustenance of the river regime, and other consumptive and non-consumptive uses…".[129] To deal with problems relating to water supply and sanitation in rural areas, to "[f]acilitate availability of safe and affordable drinking water supplies through various means" and to "[m]andate relevant public water and sewerage institutions to provide necessary drainage and sanitation … in the interest of public health" are among the government policies.[130]

In the Bhutan Water Policy 2004, the human right is explicitly recognised and the first priority of water for the human right to water is guaranteed. The Policy asserts that "[w]ater is a precious natural resource and a basic human need for survival", [131] and is aware of that "every individual has the right of access to safe and sufficient quantity of water for personal consumption and sanitation at an affordable price".[132] Further, the goal of the government is set forth as "to provide universal access to safe water and sanitation", and "[n]ational standards for water

---

124 See, *National Implementation of Agenda 21 – 2002 Country Profiles*, available at UN webpage http://www.un.org/esa/sustdev/natlinfo/cp2002.htm (last visited 31/05/2005).

125 See the relevant discussion in 5.2.3 of Chapter 5.

126 See para. 1.1, *National Water Policy 2002*, India, p.1; also available at http://wrmin.nic.in/policy/nwp2002.pdf (last visited 31/05/2005).

127 *Id.*, para. 1.8, p. 2.

128 See para. 2(a), *National Water Policy 1995*, Mozambique, available at http://www.dna.mz/natwpol.htm (last visited 31/05/2005).

129 See para. 4.3(b), *National Water Policy 1999*, Bangladesh, Ministry of Water Resources, *National Water Management Plan Project* (Vol. 5) (2001), Annex A.

130 *Id.*, para. 4.6(a) and (c).

131 See para. 3.2, *Bhutan Water Policy 2004*, available at http://www.saciwaters.org/db_bhutan_water_policy.htm (last visited 31/05/2005).

132 *Id.*, para. 5.2.1.

quality shall be developed. The best available water sources shall be allocated for drinking purposes".[133] As to the water user priorities, in the allocation of water, "[w]hen water resources is not sufficient both in quantity and quality, to meet every legitimate demand, water for drinking and sanitation for human survival shall be the primary consideration", and the general priority ranking principle shall be drinking water and sanitation, irrigation …".[134]

2.4.3 Conclusion

To sum up, the state practice is strong enough to support the human right to water. Therefore, some leading experts point out that, "it can be concluded that a number of legal instruments, as well as actions, at local, national, … levels are emerging and reinforcing the recognition of the human right to water".[135]

2.5 WATER RIGHTS: THE ENVIRONMENTAL RIGHT TO WATER

The environment[136] has its own intrinsic and natural value,[137] although the objective of environment protection may be a comprehensive or holistic

---

[133] *Id.*, para. 5.2.2.

[134] *Id.*, paras. 5.1.1 and 5.1.3.

[135] See Salman, S. M. A. & McInerney-Lankford, S. (2004), *supra* note 114, p. 83.

[136] The word environment employed here includes, (a) the non-living (abiotic) environment of air, water and land; and (b) the living (biotic) environmental of *flora, fauna* (plants and animals), it mainly refers to the latter one, and sometimes changeable with the word ecosystem, which "refers to the communities of plants, animals and other organisms, and the physical environment of any given place, these elements being linked by the flow of energy and the circulation of materials from producers to consumers and decomposers". See UNESCO, et al. (2003), *supra* note 9, p. 131.

[137] This could be found from the evolution of the ethics relating the environment since the 1970's. In the UN Conference on the Human Environment 1972, human being was recognized as "both creature and moulder of his environment", and "[m]an has a special responsibility to safeguard and wisely manage the heritage of wildlife and its habitat". See para. 1 of Preamble, and Principle 4, *Declaration of the UN Conference on the Human Environment*, 16 June 1972, UN Doc A/CONF 48/14 (1972); 11 ILM. 1972. 1416. It is obviously that the reasons for the environment protection only lay on the basis of human interest at that time.

In accordance with the *Protocol Additional to the Geneva Conventions of 12 August 1949, and relating to the Protection of Victims of International Armed Conflicts (Protocol 1)*, "[c]are shall be taken in warfare to protect the natural environment against widespread, long-term and severe damage", and "[a]ttacks against the natural environment by way of reprisals are prohibited". See art. 55, *Protocol Additional to the Geneva Conventions of 12 August 1949, and relating to the Protection of Victims of International Armed Conflicts (Protocol 1)*, 1125 UNTS 3, entry into force 7th December 1979.

According to Article 2 of the African Convention on the Conservation of Nature and Natural Resources 1968, the fundamental principle of the Convention is that, "[t]contracting States *shall undertake to adopt the measures necessary to ensure conservation, utilization and development of soil, water, flora and faunal resources in accordance with scientific principles* and with due regard to the best interests of the people" [emphasis added]. *African Convention on the Conservation of Nature and Natural Resources*, 15 September 1968, entered into force 16 May 1969, available at African Union webpage http://www.africa-union.org/Official_documents/Treaties_%20Conventions_%20Protocols/Convention_Nature%20&%20Natural_Resources.pdf (last visited 31/05/2005).

In the *UN World Charter for Nature* adopted at the 37th UN General Assembly on 28th October 1982, one principle is "[n]ature shall be respected and its essential processes shall not be impaired". See para. 1, *UN World Charter for Nature*, 28 October 1982, UN Doc A/RES/37/7; also available at http://www.un.org/documents/ga/res/37/a37r007.htm (last visited 31/05/2005). One should note that, the Declaration of the UN Conference on the Human Environment 1972 and the UN World Charter for Nature 1982 are non-binding instruments, which do not impose any direct legal obligations on the States. However, they can be considered as evidence of possible new laws that the States might be prepared to adopt as bindings.

In the UN Conference on Environment and Development 1992, many legal documents relating to environment as well as development were adopted. In the Rio Declaration on Environment and Development, the ideology of "the integral and interdependent nature of the Earth" is further understood and introduced. See para. 6 of Preamble, *Rio Declaration on Environment and Development*, 16 June 1972, UN Doc A/CONF.151/26 (Vol.I).

In the *UN Convention on Biological Diversity 1992*, the intrinsic value of biological diversity, as well as the ecological, genetic, social, economic, scientific, educational, cultural, recreational and aesthetic values of biological diversity and its components, are definitely recognised. See para. 1 of Preamble, *UN Convention on Biological Diversity*, 5 June 1992, 1760 UNTS 79, entered into force 29 December 1993; also available at http://www.biodiv.org/convention/articles.asp (last visited 31/05/2005).

According to the Declaration and Treaty of Southern Africa Development Community 1992, one of its objectives is to "achieve sustainable utilisation of natural resources and effective protection of the environment". See art. 5 (1) (g), *Declaration and Treaty of South Africa Development*, 17 August 1992, entered into force 30 September 1993, available at http://www.iucnrosa.org.zw/elisa/SADC-Protocols/declaration%20_treaty.html (last visited 31/05/2005).

In accordance with the Agreement on the Cooperation for the Sustainable Development of the Mekong River Basin, the development, utilisation, conservation and management of the water resources Mekong River Basin should be in "consistent with the needs to protect, preserve, enhance and manage the environmental and aquatic conditions and maintenance of the ecological balance exceptional to this river basin" is reaffirmed. See para. 5 of Preamble, *Agreement on the Cooperation for the Sustainable Development of the Mekong River Basin*, 5 April 1995, entered into force 5 April 1995, available at Mekong River Commission webpage

package. In other words, the environment shall be considered as a subject of law. Further, it shall be entitled to enjoy a healthy status, including obtaining appropriate water in terms of quality and quantity. After the intrinsic value and the right of the environment are examined, the legal status of the environmental right to water will be analysed.

2.5.1 The intrinsic value and the right of the environment

Indeed, the environment protection encompasses many different spheres, and various approaches have been employed. According to the studies of Gillespie, A., regarding the reasons "why the environment is protected and how these ideas are presented in the international environmental arena of policy and law", an anthropocentric justification, as well as a non-anthropocentric justification, has been deployed.[138] The anthropocentric justification consists of seven different approaches, *i.e.*, the self-interest justification, the use of economic rationale justification, the religious justification, the rights of future generation justification, the aesthetic justification, the cultural justification and the

---

http://www.mrcmekong.org/pdf/95%20Agreement.pdf (last visited 31/05/2005); also 34 ILM 864 (1995).

In the *UN Millennium Declaration*2000, respect of nature is listed among six fundamental values to be essential to international relations in the 21st century. See para. 6, *UN Millennium Declaration* 2000.

In the *Johannesburg Declaration on Sustainable Development 2002*, on the basis of reaffirming commitment to sustainable development, the thought of sustainable development consisting of economic development, social development and environmental protection is introduced, and "[t]he global environment continues to suffer" is among the challenges human beings are facing. See paras. 1, 5 and 13. *Johannesburg Declaration on Sustainable Development*, 4 September 2002, UN Doc. A/CONF.199/20.

In the area of international trade, the value of the environment is specially mentioned. For example, in the *Agreement Establishing the World Trade Organization* (WTO), the intrinsic value of the environment or the nature has been recognized, and sustainable development has been committed. See para. 1 of Preamble, *Agreement Establishing the World Trade Organization*, 19 April 1994, available at WTO webpage http://www.wto.org/english/docs_e/legal_e/04-wto.doc (last visited 31/05/2005). Further, the *WTO Doha Ministerial Declaration* 2001 indicates that, "under WTO rules no country should be prevented from taking measures for the protection of human, animal or plant life or health, or of the environment at the levels it considers appropriate … ." see para. 6, *WTO Doha Ministerial Declaration*, 14 November 2001, WT/MIN(01)/DEC/1; also available at WTO webpage http://www.wto.org/english/thewto_e/minist_e/min01_e/mindecl_e.htm (last visited 31/05/2005).

[138] See Gillespie, A. (1997) *International Environmental Law, Policy and Ethics*, Clarendon Press, Oxford, pp.1-2.

recreational justification.[139] The non-anthropocentric justification embraces three approaches, *i.e.*, the moral considerability of animals justification, the respect for life justification and the land ethic justification.[140] Each of these approaches, however, shares a common shortcoming, in that, without one or more of other approaches being employed simultaneously, it cannot resolve all issues relating to the protection of the environment.[141]

From a legal perspective, the environment protection concerns property law, human rights law, criminal law, administrative law, procedure law, etc. "At a wider level, the work of environmental philosophers may be adopted so as to establish a legal concept which accords the environment some status in its own right."[142] Regarding the employment of the term of "environmental right", one expert thinks that, "[t]he problem of defining such (environmental) rights so as to satisfy diverse ethical criteria is complicated by the need to make them operate in a legal context".[143] Others assert that the term "might be better used to describe procedural rights possessed by individuals, such as rights to information, rights to be consulted, rights to have reasons provided for a decision and rights to make a complaint".[144]

In this work, the land ethic justification will be employed to analyse the environmental right to water. On the basis of the ethic of humanity being part of nature and the environment having its intrinsic value, the author believes that it is reasonable and logical to accept the following two concepts to a certain extent: (a) the environment not only has its own intrinsic value but also provide human beings with essential services; and (b) the sustainability of the environment requires participatory, environment-based management. In this research, the author considers the environmental right as: (a) the right that the environment should have, *i.e.*, a substantive content; and (b) the right as well as the duty for the present human beings as well as groups formed by them at various levels, including individuals, communities, governments, governmental organizations, non-governmental organizations, the present generation, etc., to safeguard (a), *i.e.*, a procedural content. Therefore, in order to realise the environmental right, the present human beings should be entitled to the right as well as be given the duty to protect the environment for the benefit of

---

139 *Id.*, p. 18.
140 *Id.*, p. 135.
141 *Id.*, pp. 176-178.
142 See Bell, S. & McGillivray, D. (2000) *Environmental law* (5th ed.), Blackstone Press, London, p. 19.
143 See Anderson, M. R. (1998) Human Rights Approaches to Environmental Protection: An Overview, in *Human Rights Approaches to Environmental Protection* (ed. Boyle, A. E. and Anderson, M. R.), Clarendon Press, Oxford, p. 11.
144 See Bell, S. & McGillivray, D. (2000), *supra* note 142, p. 19.

individuals, communities, the present generation and the future generation, and for the benefit of the environment.

2.5.2 The legal status of the environmental right to water

Water is an essential part of the environment, and plays multiple environmental functions. Without water, life cannot exist. Water resources also play a vital role in the conservation of the whole environment. In this sense, "[w]ater is an essential part of any ecosystem".[145] That is to say, without a minimum amount of water of proper quality in the certain places, the environment cannot sustain itself in a good and healthy order. The environment is, therefore, entitled to receive a minimum amount water of proper quality in certain places for its beneficial order or sustainable ability.

However, unless the present human beings or groups at different levels organised by them have the right and are under the duty to protect the substantive content of the environmental right to water, it is impossible for the substantive content of the right to be realised. In this work, the environmental right to water would be invoked as, (a) the entitlement of the environment to receive a minimum amount of water of proper quality in certain places to sustain itself in a good and healthy order, *i.e.*, the substantive content of the environmental right to water, and (b) the right and duty of the present human beings or groups organised by them to realise (a), *i.e.*, the procedural content of the environmental right to water. The environmental right to water constitutes the main and most important content of the environmental right. As discussed below, water has been recognised as an indispensable component of the environment, and the environmental right to water has been recognized in many international and domestic legal documents, albeit implicitly in most cases.

On the one hand, at the international level, many comprehensive as well as special international documents dealing with or involving water resources have been adopted, and the environmental right to water has been stipulated, especially since the UN Conference on the Human Environment convened in Stockholm in 1972.[146] The relevant content of the environmental right to water encompassed in international law could be employed as an aspirational standard by the States in their legislation and policy-making.

On the other hand, the substantive content of the environmental right to water as well as the procedural content of the environmental right to water have been stipulated in many jurisdictions, although the legislations relevant to the environmental right to water vary from jurisdiction to jurisdiction because of

---

145 See UNESCO et al. (2003) *The United Nations World Water Development Report: Water for People, Water for Life (Executive Summary)*, UNESCO Publishing/Berghahn Books, Paris, p. 13.

146 See the detailed discussion in 4.3 of Chapter 4.

different water situations and diverse levels of economic and social development. Firstly, many jurisdictions recognise that the environmental requirements for water should be met as a priority, or at least it is one of beneficial uses should be taken into consideration, although the approach adopted varies from jurisdiction to jurisdiction. Secondly, regulations in many jurisdictions have empowered the relevant water management institutions or authorities to adopt measures to ensure that there is a minimum water of quality for environmental use, and have allowed the relevant stakeholders to participate in the decision-making process and decision implementing, although only the water for certain vital or important parts of the environment is regulated in detail.

Under Chinese water law, ecological-environmental water-use is among the beneficial water uses, and the protection of water resources is linked to both the ecology and the environment.[147] In South Africa, the water required to maintain environment sustainability is a right.[148] Under South Australia water law, water dependent ecosystems are protected.[149]

Under the French Water Act 1992, "respecting natural balances" is declared to be "in the general interest". [150] Article 2, which provides the purpose of the water resources management, lists: (a) "the preservation of aquatic ecosystems, of sites and wetlands; wetlands are understood as land, whether used or not, usually flooded or constantly or temporarily drenched with water, that is fresh, brackish or salt; most of the plants, if any, are hygrophilous at least part of the year"; (b) "the protection against all pollution and the restoration of surface and groundwater quality, and ocean water within the boundaries of the territorial waters"; (c) "the development and protection of water resources", and (d) "the conservation and free flow of water".[151]

The UK Water Resources Act 1991, regarding the general duties of the relevant authorities, regulates the general environmental and recreational duties, environmental duties with respect to sites of special interest, and codes of practice with respect to environmental and recreational duties.[152] Further, it provides "minimum acceptable flows".[153] In practice, six overall uses of water have been defined for the purpose of implementing the new national Statutory Water Quality Objectives scheme, two of which relate to the environmental right to water, *i.e.*, the fisheries ecosystem and the special ecosystem.

---

147 See the detailed discussion in 3.6 of Chapter 3.
148 See the detailed discussion in 5.2.4 of Chapter 5.
149 See the detailed discussion in 5.3.4 of Chapter 5.
150 See art. 1, French Water Act 1992.
151 *Id.*, art. 2.
152 See sections 16, 17 and 18, Water Resources Act 1991.
153 *Id.*, sections 21, 22 and 23.

In the Germany Act on Managing Water Resources 2000, the first principle provides that, "[a]s a part of the ecobalance and as a habitat for animals and plants, the waters are to be secured. They shall be managed in such a way that they serve the general well-being and, in harmony with this, also serve the purposes of an individual and that any avoidable damage to their ecological functions does not occur".[154]

In the USA, the Water Resources Development Act 1986 provides that, the Secretary of the Army "may carry out an aquatic ecosystem restoration and protection project if the Secretary determines that the project - (1) will improve the quality of the environment and is in the public interest …".[155]

2.6 CONCLUSION

In this chapter, the definition and utilisation of water resource have been examined. From the evolution of water rights, it is easy to understand that the development of technology, society, economic, as well as the geographical and cultural background, has been affecting the ways in which water resources are utilised, as well as the types of utilisation of water resources. Further, these have influenced the legal forms and substantive content of water rights, and thus made "water rights" a changing concept. At present, fitting the vital values of water resources, *i.e.*, the economic, social and environmental values, the legal content of water rights consists of three aspects, *i.e.*, the property right on water resources, the human right to water, and the environmental right to water. Under a good water rights mechanism, these three aspects, where proper, in all or in part, and the integration and administration of water rights, should be addressed with care.

To sum up, this chapter has developed an analytical approach which consists of four key elements, namely, the definition of water resource, the property right on water resources, the human right to water, and the environmental right to water. Following this analytical approach, the following five chapters will discuss these elements, where appropriate, wholly or partially.

---

[154] See art. 1 (a) (1), Germany Act on Managing Water Resources 2000.

[155] See § 2330 (a), Water Resources Development Act 1986, 33 U.S.C. 2201 et seq.

3

Water Rights in China: Law, Practice and Problems

In order to create a sound understanding of the necessity and urgency of establishing an appropriate water rights system in China, this chapter will firstly introduce the situation regarding water resources there. Secondly, to make it easier for western readers to understand Chinese water law, Chinese legal system, including water legal norms will be discussed in a brief way, with the focus on the tradition of transplanting components from the civil law jurisdiction. Then, applying the analytical approach developed in Chapter 2, this chapter will discuss current water rights regime in China, in particular, identify its problems. Finally, a short conclusion will be presented. In concert, this chapter discusses the water rights problems in current Chinese water law and practice, which is essential for examining the relevant aspects of water rights in international law, as well as the domestic laws in the selected jurisdictions.

ISBN: 1843391082. Published by IWA Publishing, London, UK.

3.1 WATER RESOURCES SITUATION IN CHINA

To recognise the water crises in China assists the reader to become aware of the need and importance of improving current Chinese water rights system. With a literature review, this section will provide an overview of the Chinese water crises.

3.1.1 Water resources and its utilisation[1]

In terms of total amount, with a land area of 9.6 million km^2,[2] China is rich in different types of water resources, including physical water itself, water transport resource, hydropower resources, watercourse, surface water, groundwater, etc.[3] For example, the total natural renewable water resources received by China in terms of water quantity is estimated at 2.8 thousand billion m^3 per year, and rich hydropower potential exists in China's river systems.[4] According to a national survey about hydropower potential, there are more than 11,000 sites suitable for the construction of hydropower stations with an installed capacity of over 500 kw each.[5]

However, because of a huge population of about 1.3 billion,[6] the amount per capita available is quite low in China. For example, in terms of water quantity, the amount per capita is only about 2,200 m^3 per year, which makes China one of the most water deficient countries in the world.[7] Statistics show that China

---

1 The word content of this section and the next section 3.1.2 is a revised and updated version of Part II "Water Resources and Its Utilisation" and Part III "Main Issues in Water Utilization" of article "The New Development of Water Law in China" co-authored by the author and other leading scholars, which was published in *Water Law Review*. See Wouters, P. K., et al. (2004) The New Development of Water Law in China, *Water Law Review* 7 (2) 243-307.

2 See National Bureau of Statistics of the PRC (NBS) (2003), *China Statistical Yearbook (2003) (Chinese-English version),* China Statistics Press, Beijing, p. 4.

3 See MWR, *Annual Report 2003 (English version)*, p. 3; available at MWR webpage http://www.mwr.gov.cn/english1/pdf/Annual2003.pdf (last visited 31/05/2005).

4 *Id.*

5 *Id.*

6 See NBS (2003), *supra* note 2, p. 97.

7 See Wang, S. (2001) Explanation to Water Law of the People's Republic of China (Amendment Draft) (24 December 2001), Chinese version, *Gazette of the Standing Committee of the NPC* (2002) 5, 371–374. See also MWR, *China Water Resources Gazette 2000,* its Chinese version available at MWR webpage http://www.shuiziyuan.mwr.gov.cn/gongshi/show_gazettea.asp?id=79 (last visited 31/05/2005); *China Water Resources Gazette 2001*, its Chinese version available at http://www.shuiziyuan.mwr.gov.cn/gongshi/show_gazettea.asp?id=114 (last visited 31/05/2005; and *China Water Resources Gazette 2002*, its Chinese version available at http://www.mwr.gov.cn/ztbd/zgszygb/20021231/34133.asp (last visited 31/05/2005).

sustains approximately 22 percent of the world's population, while having only 6 percent of the world's renewable water resources in terms of water quantity, and is among the world's thirteen "thirstiest" countries. Its annual average water supply per capita, which now stands at 430 m^3, is only a quarter of the world's average.[8] Additionally, the lack of potable water has created an enormous demand for bottled water, which makes China the largest bottled water market in Asia and among the top ten markets in the world.[9]

According to a document written by the Water Resources Bureau of the MWR, the disproportionate availability of water in terms of both location and time further aggravates this fact. The wet season in China typically occurs from April to September, when about 60-70 percent of the internal renewable water resources is received. However, wide fluctuations in precipitation from year to year, including frequently occurring drought years and years with heavy rainfall, have resulted in serious water disasters, such as floods and droughts. The geographic distribution of water in terms of quantity in China is also quite uneven, with far more water being available in the south than in the north. Southern China, including the area south of the Changjiang River, accounts for 53.5 percent of China's population and 34.7 percent of its total land territory, but the South's water resources amounts to 81 percent of the country's total. Conversely, Northern China, including the river basins of the Liaohe River, Hailuanhe River, Huanghe River and Huaihe River, has 38.8 percent of the country's population and 45.2 percent of its total land territory, while its water resources merely amounts to 9.6 percent of China's total. A specific example of this geographic disparity is the Hailuanhe River Basin, which annually receives only 357 m^3 of water per person, approximately 17 percent of the country's average.[10]

One illustration is the situation that arose in 1997. Sixty percent of the surface water in the Hailunhe River Basin was diverted for irrigation, industry and domestic uses; 91.4 percent of the surface water in the Huanghe River Basin and 80.2 percent of the surface water in the Huaihe River Basin was diverted

---

[8] See Xinhua News Agency, *Water Saving: A Hot Topic in China*, 22 March 2002, available at http://www.xinhuanet.com/english/index.htm (last visited 31/05/2005).

[9] See Beverage Marketing Corporation (2003), *2003 GLOBAL BOTTLED WATER REPORT: A WORLDVIEW*, p. 4, 208.

[10] See Water Resources Department of MWR (1999) Achievements and Prospect of Development and Utilisation of Water Resources in China, *China Water Resources* (1999) 5, p. 11-12.

respectively.[11] Groundwater was removed to augment surface diversions, and this accounted for 95.7 percent, 31.0 percent, and 33.9 percent of the total shallow groundwater, respectively, in each of the above river basins.[12]

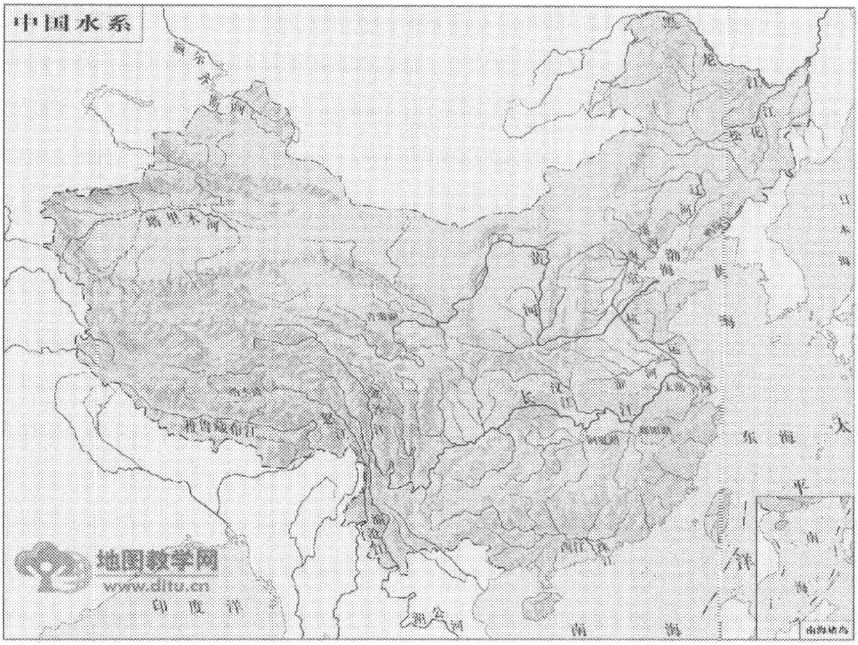

Figure 3.1. Main rivers and lakes in China (Maps Education Web of Sino Maps Press webpage http://www.ditu.com.cn/fw/main_fw_map_11.htm).

In China, the three main uses for surface water are irrigation, industry, and domestic use. In 2000, the total water used (excluding the eco-environmental use) was 549.8 billion m^3, of which 63 percent was for irrigation, 20.7 percent for industry, 10.5 percent for domestic purposes, and 5.8 percent for forestry, fisheries and animal husbandry.[13] In 2001, the total water used (excluding that for the eco-environmental use) was 556.7 billion m^3, of which 62.6 percent was for agricultural irrigation, 20.5 percent for industry, 10.8 percent for domestic

[11] See MWR, *China Water Resources Gazette 1997*, its Chinese version available at MWR webpage http://www.shuiziyuan.mwr.gov.cn/gongshi/show_gazettea.asp?id=3 (last visited 31/05/2005).

[12] *Id.*

[13] See MWR, *China Water Resources Gazette 2000*.

purposes, and 6.1 percent for forestry, fisheries and animal husbandry.[14] It is clear that agricultural irrigation is the largest water user in China.

3.1.2 Main issues in water utilisation

Water shortage, water pollution, and the potential for natural water disasters are the three primary problems related to water resources management in China, and each has negatively impacted on Chinese economic and social development as well as environmental improvement. [15] The floods within the major river basins in China have attracted world attention, especially the flooding in the Changjiang River in 1998 and the extraordinary floods of the Huanghe River in 1933 and 1958.[16] The water scarcity problems associated with drought exist mainly in the north and northwest parts of China, but have now extended into the south China.[17] Being the semi-arid or arid regions in China, the north and northwest lack water, and the environment needs precipitation for its preservation.

The problems that water shortages cause for economic development tend to be even more serious. At present, over 400 of the 668 cities in China suffer water shortages, producing an annual water supply deficit of more than 6 billion m^3.[18] One hundred of these cities, including Beijing, Tianjing, Xi'an, Taiyuan, Datong, Qingdao, Yantai and Dalian, suffer serious water shortages[19]. The shortages in these areas have driven the over-abstraction of groundwater to meet the water supply deficit, which has resulted in greatly reduced local water tables, ground surface subsidence and seawater infiltration in the coastal areas.[20]

These water supply shortages also greatly contribute to environmental degradation. For example, the zero flow of water of the Huanghe River occurred initially in the 1970s and has gradually worsened year by year since that time.

---

[14] See MWR, *China Water Resources Gazette 2001.*

[15] See Zhang, G., et al. (2000) The Comprehensive Report on Strategy of Water Resources of Sustainable Development in China, *China Water Resources* (2000) 8, p.7.

[16] See Hu, M. & Luo, C. (1989) The Ministry of Water Resource and its Power, *The Historical Extreme Floods in China* (1989) 1, pp. 171-187.

[17] See MWR, *China Water Resources Statistics Gazette 2002*, its Chinese version available at MWR webpage http://www.mwr.gov.cn/gonggao/030613.doc (last visited 31/05/2005).

[18] See Wang, S. (2001), *supra* note 7.

[19] See Jiang, W. (2000) Research on Water Resources Safety of China in the 21st Century, *China Water Resources* (2000) 8, p. 41.

[20] See MWR, *China Water Resources Gazette 2000*, *China Water Resources Gazette 2001*, and *China Water Resources Gazette 2002*.

The most serious zero flow recorded to date was in 1997, with the longest extent of zero flow of water -- 700 km, amounting to 90% of its length downstream, and with the longest period -- 226 days in a year.[21] Many institutions and experts continue to research the issues that have resulted from these instances of zero flow of water.[22] The Tarim River, the most important river in the northwest China, also has a history of zero flow of water for approximately 30 years. Its complete length of zero flow of water is more than 320 km downstream, resulting in the drying up of the Lop Nur Lake, serious ecosystem degradation, and desertification.[23] All of these are threatening the very existence of the indigenous environment.[24] Indeed, the issues of the Tarim River appear to be even more serious than those of the Huanghe River.

In 2000, China experienced a more severe situation.[25] Although the total water used accounted for only 19.8 percent of the total water resources in China, most of northern China and some of southern China experienced the most severe drought since 1949. The drought affected over 20 provinces, ethnic autonomous regions or municipalities directly under the Central Government (CGMs).[26] The drought affected agricultural lands in an area of nearly 41 million hectares, and approximately 27.7 million people in rural areas and 17 million large livestock (such as horses, oxen, cows, donkeys, and mules) suffered from shortages of drinking water.[27] In 18 provinces, ethnic autonomous regions or CGMs, 620 cities experienced water shortages that affected more than 26 million of the urban population.[28]

One of measures adopted by the Chinese government to resolve the water shortages in certain areas is inter-basin water transfer. Many projects for inter-basin water transfer have been instigated or built, and the eastern and middle

---

[21] See Jiang, W. (2000), *supra* note 19, p. 11.

[22] See Ye, Q. (1998) Flow Interruptions and Their Environmental Impacts on the Yellow River Delta, *ACTA GEOGRAPHY SINICA* (1998) 5, p. 385-391.

[23] See Zhu, X. (2001) Adapting New Measures to Strengthen Comprehensive Rehabilitation in Tarim River Basin in the New Situation, *China Water Resources* (2001) 9, pp. 50-52.

[24] See Fan, Z., et al. (1998) Study on the Ecological Environment Evolution and Sustainable Development in Tarim River Basin, in *Proceedings of Seminar on the Water Resources, Environment and Management of Tarim River Watershed (1998)*, pp. 171-177.

[25] See MWR, *China Water Resources Gazette 2000*.

[26] China is administratively divided into 23 provinces, 5 ethnic autonomous regions, 4 CGMs and 2 special administrative regions. An ethnic autonomous region or a CGM has the same political, economical and jurisdictional rights as a province.

[27] See MWR, *China Water Resources Statistics Gazette 2000*, its Chinese version available at MWR webpage http://www.cws.net.cn/gazette/tongjigb/index.html (last visited 31/05/2005).

[28] *Id.*

routes of the South-to-North Water Diversion Project are under construction.[29] By the end of 2003, the inter-basin water transfer projects were able to transfer water 15.3 million m^3 annually.[30]

Figure 3.2. Administrative regions at the provincial level in China (Xinhua News Agency webpage http://news.xinhuanet.com/english/2003-02/19/content_815536.htm).

Water pollution is also a serious problem in China. Most rivers and lakes, and groundwater in some cities, suffer from pollution.[31] More precisely, nearly 50 percent of river length and 90 percent of city water supply are polluted to differing extents.[32] Thus, water pollution prevention has been widely implemented by introducing uncompromising measures in China, especially in

---

[29] See, MWR, *Annual Report 2003 (English version)*, p. 4.

[30] *Id.*

[31] See MWR, *China Water Resources Gazette 1997*.

[32] See Wang, S. (2001), *supra* note 7.

the three most seriously polluted rivers (*i.e.*, the Huaihe River, Haihe River, and Liaohe River) and the most three seriously polluted lakes (*i.e.*, the Taihu Lake, Dianchi Lake, and Chaohu Lake).[33]

Many factors, including engineering practice and certain non-engineering activities, have caused these water issues in China.[34] Water resources management and related policies and legislation are obviously one of the main aspects that affect these issues.[35] Considering that water resources management is the most important factor in achieving rational water utilisation and sustainable development in developing countries, especially in rapidly developing countries such as China, it is important to analyse the present legal and institutional system, and to ensure that it is able to adapt to changing situations and meet the water demands of the future.

3.2 CHINESE LEGAL SYSTEM: AN OVERVIEW

In this section, a literature approach will be used to trace the evolution of the Chinese legal system, to provide the background of Chinese law and, in particular, to examine the influence of the civil law system. Then, the legislative bodies and the hierarchy of Chinese law will be introduced in accordance with the relevant legal provisions, and an outline of the current Chinese legal system and the Chinese water legal norms will be presented, for both provide necessary information for a useful discussion of water rights in China.

3.2.1 Evolution of Chinese legal system

As early as the 21st century B.C., the traditional Chinese legal system came into beings, and then continued to develop for the ensuring 4,000 years till the late 19th century.[36] However, this legal system was developed in accordance with an ideology exemplified by the well-known proverb, *i.e.*, "no land does not belong to the king, no man is not the servant of the king", and the law was deployed merely as a tool for the emperor/empress to rule the country for his/her own benefit. Such law, in most situations, referred to penalties and was completely enacted or originated by the ruler, who said to be the "Son of the

[33] See MWR, *China Water Resources Gazette 2000*, *China Water Resources Gazette 2001*, and *China Water Resources Gazette 2001*.

[34] See Zhang, D., et al. (2000), *supra* note 15, p. 16.

[35] *Id.*

[36] See *China Encyclopedia (Law Volume)* (Chinese version), p. 762.

God". Therefore, it is not correct or proper to consider that the traditional Chinese legal system was secular, as some writers do.[37]

The Opium Wars[38] and the following unequal treaties signed by the government of the Qing Dynasty with the western powers from 1840 to the end of 19th century revealed increasing corruption and incapability on the part of that government of the Qing Dynasty. The invasion of the western powers in military as well as in economic and cultural terms led to great changes in various fields in many regions of China. Gradually, the developing western legal culture and system began to come to the attention of the Qing government as well as some advanced Chinese intellectuals. Due to pressure from the western powers, requests from the Chinese intellectuals, and the continuous rebellion of the people, the Qing government reluctantly decided that it might be in a better position to stabilise its rule under the traditional legal system by learning from the west. In 1902, Shen Jiaben and Wu Tingfang were nominated as Ministers for Law Revision.[39] In the following years, many codes and writings relating to modern legal systems and culture in the west were translated, while selected persons were sent abroad by the government to investigate and/or learn legal systems in the west, while yet others travelled abroad at their own expense.[40] Subsequently, the government declared that it was "preparing a plan for the establishment of a constitutional regime" in 1906, and in August 1908, the Emperor-determined Outline of the Constitution was promulgated under the name of the emperor.[41] During the same period, many codes were drafted, *e.g.*, the Qing Criminal Code, the Qing Civil Code, the Qing Commercial Code, and so forth.[42] These laws were transplanted from the west, especially from Germany via Japan.[43] However, due to the establishment of the Republic of

---

[37] For example, Xin, C. (1999) thinks, "it (traditional China legal system) was a secular rather than a religious legal system". Xin, C. (1999) *Chinese Legal System & Current Legal Reform*, (China) Law Press, Beijing, p. 2.

[38] The writers in China and the west believe that the Opium Wars is the event that marked the beginning of a new stage of China history, and triggered the modernisation of the Chinese legal system. See Xin, C. (1999), *id.*, p 2; Yan, W. (2001) *Chinese Legal Reform*, Routledge, London, p. 9; Potter, P. B. (2001) *The Chinese Legal System*, Routledge, London, p. 8.

[39] See Xin, C. (1999), *supra* note 37, p. 6.

[40] *Id.*, p. 7.

[41] *Id.*

[42] See Yan, W. (2001), *supra* note 38, p. 10; Liang, H. (1993) *Civil Law Theories: Cases and Legislations*, (China) Law Press, Beijing, p. 60; Huang, P. C. (2001) *Code, Custom and Legal Practice in China*, Stanford University Press, Stanford, p. 26.

[43] *Id.*

China on 1st January 1912, these draft laws had no chance to be enacted and did not come into force.

Under the rule of the Republic of China on China (January 1912 – September 1949), the modernisation of the Chinese legal system continued, on the basis of the work done by the government of the Qing Dynasty.[44] Many laws were drafted and enacted by transplanting components from the western legal system, especially from Germany via Japan. The Civil Code, which was enacted in 1941, may be taken as example. Both its content and form were obviously influenced by the relevant civil law legislative experiences in Germany, Japan and Switzerland, while with the Russia Civil Code and the Thai Civil Code as references.[45]

On 1st October 1949, the Communist Party of China (CPC) led by Mao Zedong declared that the People's Republic of China (PRC) had been established, and the PRC has ruled the mainland of China from that time onwards. One of the most important things the PRC did after its foundation was to abolish all laws enacted by the Republic of China.[46] Unfortunately, the Cold War prevented Mao Zedong from developing the economy with the help of the west,[47] so the PRC had no choice but to develop and strengthen its relationship with the Union of Soviet Socialist Republics (USSR). From 1949 to 1956, China was influenced greatly in many areas by the USSR, which had a strong civil law tradition.[48] However, following the completion of the transition from the stage of the "New Democracy Revolution" to the stage of the "Socialist Revolution" in 1958, the interruption of the relationship with the USSR, and the deification of Mao Zedong, the existing laws became very limited value to the rule of the CPC, because other and more effective tools, *e.g.*, socialist public ownership, planned economy and people democracy dictatorship,[49] began to be employed.[50]

---

[44] See Liang, H. (1993), *supra* note 42, p. 63; Chen, Y. (2002) Civil Law Development: China and Taiwan, *Stanford Journal of East Asia Affairs* (2002) 2, p. 9.

[45] See Liang, H. (1993), *supra* note 42, p. 63.

[46] See Xin, C. (1999), *supra* note 37, p. 11; Chen. Y. (2002), *supra* note 44.

[47] Early in 1945, developing economic relations, e.g., trade, investment, etc., with each state was among the CPC's foreign policies. See Mao, Zedong (1945) United Government, in *Selected Works of Mao Zedong*, Vol. III, pp. 1029-1100, (China) People's Publishing House, Beijing, 1991.

[48] See Barnett, A. D. (1981) *China's Economy in Global Perspective*, The Brookings Institution, Washington D.C., pp. 2-3.

[49] "People democracy dictatorship" was first brought forward theoretically by Mao Zedong on 30 June 1949 in his article *People Democracy Dictatorship* (in Selected Works of Mao Zedong, Vol. IV, pp. 1463-1482, People's Publishing House, Beijing, 1991). According to the explanation in the CPC documents, *e.g.*, *Key Points of Talks in Wuchang, Shenzhen, Zhuhai, Shanghai etc* (Selected Works of

Although the CPC declared a socialist society to have been established, the majority of people in rural areas did not live happily life because of the lack of a basic food supply.[51] The death of Mao in September 1976 was like the loss of a deity in China, and a new deity was anticipated. On the one hand, popular uncertainty implied a possible crisis or revolution. To avoid dying of hunger, in 1977, the peasants in some regions began to break away from the official planned economy, which produced virtually a little food for them to live on.[52] On the other hand, there was a struggle for power within the CPC. Finally, Deng Xiaoping emerged as the winner, and adopted a series of measures to stabilize his power. The most important of these was the implementation of an economic policy of "reforming the domestic economy and opening to the outside".[53]

Alongside constant economic development, Deng was forced to revise or develop his economic policy. For example, regarding the aim of the economic reform, the following objectives were employed in chronological order, *i.e.*, reforming the planned economy (December 1978), the Chinese characteristic planned economy in a dominant position with market adjustment as a supplement (September 1982), the Chinese characteristic socialist planned commodity system (October 1984), the Chinese characteristic socialist planned market system (June 1989), and the Chinese characteristic socialist market system (October 1992).[54] The term "Chinese characteristic", by its very nature, is employed to emphasise the control of the CPC or the government it lead to rule the country. Law is regarded as one of the instruments for such rule.

However, due to the abuses connected to dictatorship, more and more Chinese people begin to recognise their natural rights, and the CPC and the PRC have been enacting policies and laws to adjust to a continuous developed commodity or market economy. Modernisation of the legal system had to be undertaken at least in form. Since no law relevant to a commodity or market economy existed in the traditional Chinese legal system (in words, there was no tradition of modern legal culture), it was easier, on the one hand, for the PRC to

---

Deng Xiaoping, Vol. III, pp. 370-383, People's Publishing House, Beijing, 1993), it means "democracy exists among the people, and the people dictate the enemy".

50 See Xin, C. (1999), *supra* note 37, pp. 13-18.

51 See Perkins, D. and Yusuf, S. (1984) *Rural Development in China*, The Johns Hopkins University Press, Baltimore, p. 5.

52 *Id.*, p. 80.

53 See, *Gazette for the 3rd Plenary Meeting of the 13th Central Committee of the CPC*, adopted on 22nd December 1978, available at http://www.people.com.cn/GB/shizheng/252/5089/5103/5205/20010428/454803.html (last visited 31/05/2005).

54 See Jiang, Zhemin (1992) *Report on the Fourteenth National Representatives Congress of CPC*, Legal Daily, 21 October 1992.

enact laws by learning from other developed countries. On the other hand, because it is more convenient to enact and enforce written law, and because of the modernisation history of the past, a civil law system is generally adopted, although some mechanisms employed in the common law system are deployed simultaneously, for example, the concept of trust.[55]

3.2.2 Legislative bodies and hierarchy of laws[56]

The term "law" is usually invoked in two senses in China: (a) in a general or broad sense, as a legal norm enacted or admitted by the state, and (b) in a special or narrow sense, as a written normative document adopted by the NPC or its Standing Committee.[57] However, sometimes its precise meaning should be understood in accordance with the circumstances in which it is employed. The Chinese legislative bodies and hierarchy of laws are stipulated by The Constitution of the PRC, which was promulgated in 1982 and amended four times, with a total of thirty-one amendments.[58] Other laws provide more detail in some aspects, such as the Legislation Law of the PRC,[59] and the Law of the PRC on the Organisation of Local People's Congresses and Local Governments

---

[55] For example, the Trust Law of the PRC was adopted at 21st Session of the Standing Committee of the Ninth NPC, and has been implemented since 1st October 2001. Trust Law of the PRC, official Chinese version, *Gazette of the Standing Committee of the NPC* (2001) 4, 311-316.

[56] The main content of this section and the section 3.2.4 have been published as Part IV "The Legal and Regulatory Framework in General" of article "The New Development of Water Law in China" co-authored by the author and other leading scholars, which was published in *Water Law Review*. A little revision has been made in this work. See Wouters, P. K., et al. (2004) The New Development of Water Law in China, *Water Law Review* 7 (2) 243-307.

[57] See, *Law Dictionary* (2nd ed.), Shanghai Dictionary Publisher, Shanghai 1984, p. 600.

[58] The Constitution of the PRC (2004) (hereinafter The Constitution of the PRC), official Chinese version, *Gazette of the Standing Committee of the NPC* (Special Issue, 2004), 2-27; an unofficial English version available at http://english.peopledaily.com.cn/constitution/constitution.html (last visited 31/05/2005).

[59] The Legislation Law of the PRC 2000, adopted by the 3rd Session of the Ninth NPC on March 15, 2000, entered into force 1st July 2000. Legislation Law of the PRC 2000, official Chinese version, *Gazette of the Standing Committee of the NPC* (2000) 2, 112-127. In accordance with Article 2, "[t]he enactment, amendment and repeal of any national law, administrative regulations, local regulations, autonomous regulations and special regulations shall be governed by this Law. The enactment, amendment and repeal of departmental rules promulgated by agencies under the State Council and local rules promulgated by local governments shall be carried out in accordance with the relevant provisions of this Law".

at Various Levels.[60] However, the PRC legislative structure differs from that of the USA and other parliamentary systems in the west, and the Standing Committee (discussed below) has lawmaking power when the larger legislative body is not in session.

The Constitution The Constitution is the highest law in the Chinese hierarchy of law. It is recognised as the mother law of all other laws, and forms the foundation for and has a higher force than all other laws.[61] The NPC holds the exclusive authority to amend The Constitution.[62]

Basic laws The NPC enacts and amends basic laws governing criminal offences, civil affairs, state organs and other matters.[63] However, when the NPC is not in session, its permanent body, the Standing Committee, exercises the function and power to partially supplement and amend laws enacted by the NPC provided that the basic principles of these laws are not contravened.[64]

Laws other than basic laws The Standing Committee of the NPC exercises the function and power to enact and amend laws, with the exception of those that should be enacted by the NPC.[65]

Regulations Three kinds of regulations exist: administrative regulations, local regulations, and autonomous regulations and separate regulations. (a) Administrative regulations are enacted by the State Council in accordance with The Constitution and the law.[66] The State Council, *i.e.*, the Central People's Government of the PRC, is the executive body of the NPC as well as the highest organ of state administration.[67] Administrative regulations shall be applied nationwide, except when otherwise provided by The Constitution, laws, or the particular administrative regulations itself.

---

[60] The Law of the PRC on the Organisation of Local People's Congresses and Local Governments at Various Levels, adopted by the 2nd Session of the Fifth NPC on 1st July 1979, and has been revised four times, respectively on 10 December 1982, 2 December 1986, 28 February 1995, and 27 October 2004. Law of the PRC on the Organisation of Local People's Congresses and Local Governments at Various Levels (4th Revision), official Chinese Version, *Gazette of the Standing Committee of the NPC* (2004) 7, 659-669.

[61] See para. 13 of Preamble, and art. 5 (3), (4), (5), The Constitution of the PRC.

[62] *Id.*, art. 62 (1).

[63] *Id.*, art. 62 (3).

[64] *Id.*, art. 67 (3); also art. 7 (2) and (3), Legislation Law of the PRC.

[65] See art. 67 (2), The Constitution of the PRC; also art. 7 (3), Legislation Law of the PRC.

[66] See art. 89 (1), The Constitution of the PRC; also art. 56 (1), Legislation Law of the PRC.

[67] See art. 85, The Constitution of the PRC.

(b) Local regulations are adopted by a people's congress at the provincial level or its standing committee in light of the specific situations in and actual needs of the jurisdiction.[68] In addition, in light of the specific situations in and actual needs of the jurisdiction, the people's congress and its standing committee of a major city[69] may also enact local regulations.[70]

(c) Autonomous regulations and separate regulations are enacted by a people's congresses of an ethnic autonomous region in exercising its autonomy and in light of the political, economic and cultural characteristics of the ethnicity of or ethnicities in the areas concerned.[71] Autonomous regulations or separate regulations enacted by an autonomous region become effective after review and approval by the Standing Committee of the NPC, and those enacted by an autonomous prefecture or autonomous county shall come into force after they have been reviewed and approved by the standing committee of the relevant people's congress at the provincial level.[72]

Rules There are two kinds of rules: departmental rules and local rules. (a) Departmental rules are issued by the People's Bank of China, the Auditing Agency, or any of the working departments under the State Council, *e.g.*, MWR. These organs exercise a regulatory function within the scope of their authority in accordance with national law and administrative regulations, as well as decisions and orders of the State Council.[73] A matter on which a departmental rule is enacted has to be a matter that lies within the scope of implementing national law, administrative regulations, and decisions or orders issued by the State Council.[74] Departmental rules must be applied nationwide, except when otherwise provided by The Constitution, laws, administrative regulations or the particular rules itself.

---

[68] See arts. 100, 115, The Constitution of the PRC; also art. 63 (1), Legislation Law of the PRC.

[69] In accordance with art. 63 (4) of Legislation Law of the PRC, "a major city refers to a city where the People's Government of the province or autonomous region is seated, the city where a special economic zone is located, and any other major city approved by the State Council".

[70] See art. 63 (2), Legislation Law of the PRC.

[71] See art. 116, The Constitution of the PRC; also art. 66 (1), Legislation Law of the PRC.

[72] See art. 66, Legislation Law of the PRC.

[73] See art. 90 (2), The Constitution of the PRC; also art. 71 (1), Legislation Law of the PRC.

[74] See art. 71 (2), Legislation Law of the PRC.

(b) Local rules are formulated by the people's governments at the provincial level, or those of major cities, in accordance with The Constitution, laws, administrative regulations and local regulations that are at the provincial level.[75]

The hierarchy of the above laws may be illustrated briefly in Figure 3.3.

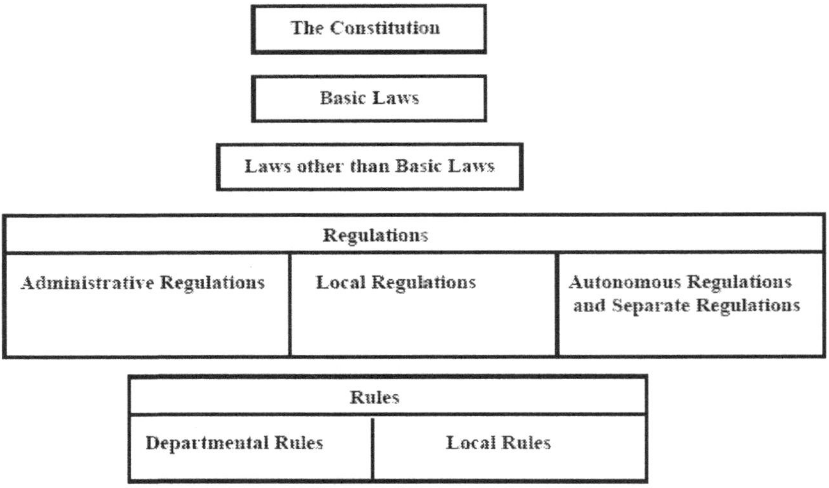

Figure 3.3. Hierarchy of Chinese Laws

3.2.3 Current Chinese legal system

Alongside the implementation of the economic policy of "reforming the domestic economy and opening to the outside" in the late 1970's, the CPC and the Chinese government have acknowledged the role of law with the slogan "To administer the State with law, to develop a socialist State under the rule of law", and the goal of building up a socialist legal system with the Chinese character by 2010.[76] The Standing Committee of the NPC has primarily categorised the Chinese legal system into seven legal branches, namely, the Constitution and

---

[75] See art. 60, Law of the PRC on the Organisation of Local People's Congresses and Local Governments at Various Levels (3rd Revision); also art. 73 (1), Legislation Law of the PRC.

[76] See Jiang Zhemin (1997) *Report on the Fifteenth National Representatives Congress of CPC*, People's Publishing House, 1997; see also art. 13, Amendment to the Constitution of the PRC (1999), official Chinese version, *Gazette of the Standing Committee of the NPC* (1999) 2, 93-94; art. 5 (1), The Constitution of the PRC.

constitutional law, civil law and commercial law, administrative law, economic law, social law, criminal law, and procedure law.[77]

The Constitution and constitutional law is the leading branch of the legal system. This branch regulates the social system, the State system, citizen's rights and duties, and the organisation, responsibility and the action principles of the State organs and authorities.[78]

The civil law and commercial law are those fundamental laws that regulate the civil and commercial activities. The laws in this branch are mainly employed to regulate the property and personal relationships between the equal legal subjects. Laws concerning property, intellectual property, debt, marriage and family, company, securities, insurance and banking are included in this branch.[79]

The administrative law regulates the State's administrative activities. It is primarily used to regulate the power of and action by the executive authorities.[80] Laws relating to administrative licensing, administrative re-consideration, etc., may be included in this branch.[81]

The economic law is deployed to adjust the social and economic relationships arising from the State administration on economic activities. The laws relating to competition, market order, the macro adjustment by the State, international economic cooperation, and the utilisation and protection of natural resources (including land, forest, grassland, water resources, etc.) are incorporated in this branch.[82]

The social law is used to adjust employment, social security and social welfare; the criminal law regulates the crime and penalties; and the procedure law regulates the procedure of litigation and arbitration.[83]

These branches are interactive and interdependent, however, and make the legal system as an organic, comprehensive and unified whole.

3.2.4 Water legal norms

Many legal norms relating to water, which will be listed or discussed below, have been enacted in China since 1979. They may be found in two kinds of

[77] See Li Peng (2001) Report 0n the Work of the Standing Committee of the National People's Congress (2001), *Gazette of the Standing Committee of the NPC* (2001) 3, 240-247. See also Li Peng (2003) Report 0n the Work of the Standing Committee of the National People's Congress (2003), *Gazette of the Standing Committee of the NPC* (2003) 2, 196-204.

[78] *Id.*

[79] *Id.*

[80] *Id.*

[81] *Id.*

[82] *Id.*

[83] *Id.*

laws: (a) special laws pertinent to water, and (b) other laws in which a provision or provisions relevant to water is/are included. As to the latter, Article 9 of The Constitution, which provides the basic system, *e.g.*, the ownership, utilisation, protection, etc, of natural resources including water resources,[84] should be mentioned specifically for its fundamental characteristics in relation to other water legal norms.

As to the special laws pertinent to water, a system has been constructed. At the level of laws other than basic laws, these laws include:

-- the Water Law of the PRC (Revision), *i.e.*, the China Water Law 2002;[85]
-- the Law for Water Pollution Control of the PRC (Revision);[86]
-- the Flood Control Law of the PRC;[87]
-- the Law on Water and Soil Conservation of the PRC.[88]

---

84 Article 9 of The Constitution of the PRC reads as follows:

> "All mineral resources, waters, forests, mountains, grassland, un-reclaimed land, beaches and other natural resources are owned by the state, that is, by the whole people, with the exception of the forests, mountains, grasslands, un-reclaimed land and beaches that are owned by collective in accordance with the law.
>
> The state ensures the rational use of natural use of natural resources and protects rare animals and plants. Appropriation or damaging of natural resources by any organisation of individual by whatever means is prohibited."

The Constitution of the People's Republic of China (1999) (English version), (China) Law Press, 2001. As to "waters" in paragraph one, which is translated from Chinese expression "shui liu", the author thinks the expression "flows of water" should be more accurate in term of literal translation.

85 The Water Law of the PRC was primarily adopted at the 24th Session of the Standing Committee of the Sixth NPC on 21 January 1988, entered into force 1st July 1988 (hereinafter "China Water Law 1988"). China Water Law 1988, official Chinese version, *Gazette of the Standing Committee of the NPC* (1988) 1, 4 -11. On 1st October 2002, it was repealed by the China Water Law 2002.

86 The Law for Water Pollution Control of the PRC was primarily adopted at the Fifth Session of the Standing Committee of the Sixth NPC on 11 May 1984, and came into force on 1 November 1984. On 15 May 1996, it was amended at the 19th Session of the Standing Committee of the Eighth NPC, and the amendment came in force on the same day. Law for Water Pollution Control of the PRC (Revision), official Chinese version, *Gazette of the Standing Committee of the NPC* (1996) 4. 34 -44.

87 The Flood Control Law of the PRC was adopted at the 24th Session of the Standing Committee of the Sixth NPC on 29 August 1997, and came into force on 1st January 1998. Flood Control Law of the PRC, official Chinese version, *Gazette of the Standing Committee of the NPC* (1997) 5, 597-608.

88 The Law on Water and Soil Conservation of the PRC was adopted at the 20th Session of the Standing Committee of the Seventh NPC on 29 June 1991, and came into

The most related laws include,
-- the Environment Protection Law of the PRC;[89]
-- the Environmental Impact Assessment Law of the PRC;[90]
-- the Fishery Law of the PRC (Revision);[91]
-- the Forestry Law of the PRC (Revision);[92]
-- the Grassland Law of the PRC (Revision);[93]
-- the Law for Electricity Power of the PRC.[94]

At the level of administrative regulations, twenties pieces of regulations were adopted or formulated by the State Council by the end of November 2001.[95] At present, the most important administrative regulations include:

---

force on the same day. Law on Water and Soil Conservation of the PRC, official Chinese version, *Gazette of the Standing Committee of the NPC* (1991) 4.

89 The Environment Protection Law of the PRC was adopted at the 11th Session of the Standing Committee of the Seventh NPC on 26 December 1989, and came into force on the same day. Environment Protection Law of the PRC, official Chinese version, *Gazette of the Standing Committee of the NPC* (1989) 6.

90 The Environmental Impact Assessment Law of the PRC was adopted at the 30th Session of the Standing Committee of the Ninth NPC on 28 October 2002, and came into force on 1 September 2003. Environmental Impact Assessment Law of the PRC, official Chinese version, *Gazette of the Standing Committee of the NPC* (2002) 6, 489-493.

91 The Fishery Law of the PRC was primarily adopted on 20 January 1986 at the 14th Session of the Standing Committee of the Sixth NPC, and came into force on 1st July 1986. On 31 October 2000, it was amended at the 18th Session of the Standing Committee of the Ninth NPC, and the amendment came in force on 1st December. Fishery Law of the PRC (Revision), official Chinese version, *Gazette of the Standing Committee of the NPC* (2000) 6, 608-615.

92 The Forestry Law of the PRC was primarily adopted at the 7th Session of the Standing Committee of the Sixth NPC on 20 September 1984, and came into force on 1 January 1985. On 29 April 1998, it was amended at the 2nd Session of the Standing Committee of the Ninth NPC, and the amendment came in force on the same day. Forestry Law of the PRC (Revision), official Chinese version, *Gazette of the Standing Committee of the NPC* (1998) 2, 187-194.

93 The Grassland Law of the PRC was primarily adopted at the 11th Session of the Standing Committee of the Sixth NPC on 18 June 1985, and came into force on 1 October 1985. On 28 December 2002, it was amended at 31st Session of the Standing Committee of the Ninth NPC, and the amendment came in force on 1st March 2003. Grassland Law of the PRC (Revision), official Chinese version, *Gazette of the Standing Committee of the NPC* (2003) 1, 43-50.

94 The Law for Electricity Power of the PRC was adopted at the 17th Session of the Standing Committee of the Eighth NPC on 28 December 1995, and came into force on 1st April 1996. Law for Electricity Power of the PRC, official Chinese version, *Gazette of the Standing Committee of the NPC* (1995) 8, 3-13.

-- the Measures on Implementation of Water-Drawing Licence System;[96]
-- the Regulations for River Channel Management;[97]
-- the Regulations for Flood Prevention;[98]
-- the Regulations for Soil Conservation;[99]
-- the Regulations for Compensation for Land Occupation and Resident Resettlement in Construction of Large-and-medium-sized Water resources and Hydropower Projects.[100]

However, the Water Industry Policy,[101] approved by the State Council, issued by the State Planning Commission on September 4, 1997, shall also be considered as part of the administrative regulations, not only because it has an equivalent legal effect as other administrative regulations, but also because it has played an important role in water resources administration in China.

At the level of departmental rules, the MWR, where appropriate, in collaboration with other ministries or committees under the State Council, had issued 90 pieces of rules to implement the laws and regulations by December

---

95 Jing, Z. (2002) *Speech in the Press Conference on the Promulgation and Implementation of New Water Law*, 5 September 2002, available at MWR webpage http://www.mwr.gov.cn/20020906/slxw/slxw9161.htm (last visited 31/05/2005).

96 The Measures on Implementation of Water-Drawing Licence System was adopted on 11 June 1993, and came into force on 1st September 1993; its Chinese version available at MWR webpage http://www.mwr.gov.cn/zcfg/Detail.wct?SelectID=4275&RecID=11 (last visited 31/05/2005). It should be noted that, although it was adopted in accordance with the Article 32 of the China Water Law 1988, it is still valid.

97 The Regulations for River Channel Management was adopted by the State Council on 3rd June 1988, promulgated and came into force on 19 June 1988; its Chinese version available at MWR webpage http://www.mwr.gov.cn/zcfg/Detail.wct?SelectID=4275&RecID=20 (last visited 31/05/2005).

98 The Regulations for Flood Prevention was adopted by the State Council on 28 June 1991, promulgated and came into force on 2nd July 1991; its Chinese version available at MWR webpage http://www.mwr.gov.cn/zcfg/Detail.wct?SelectID=1816&RecID=15(last visited 31/05/2005).

99 The Regulations for Soil Conservation was enacted by the State Council on 1st August 1993, and came into force on the same day; its Chinese version available at MWR webpage http://www.mwr.gov.cn/zcfg/Detail.wct?SelectID=1292&RecID=12(last visited 31/05/2005).

100 The Regulations for Compensation for Land Occupation and Resident Resettlement in Construction of Large-and-medium-sized Water resources and Hydropower Projects was promulgated by the State Council on 15 January 1991, and came into force on 1st May 1991; its Chinese version available at MWR webpage http://www.mwr.gov.cn/zcfg/Detail.wct?SelectID=1292&RecID=17(last visited 31/05/2005).

101 The Water Industry Policy, official Chinese version, *Gazette of the State Council of the PRC* (1997) 35, 1526-1531.

2001.[102] Among others, those most relevant to water rights include the Implementation Rules on Water Industry Policy,[103] the Procedure Provisions on Inspection and Approval of Application for Water-Drawing Licence,[104] the Guidelines on the Water Rights Transfer Concerning the Mainstream of the Yellow River within Inner Mongolia and Ningxia,[105] the Opinions on Water Rights Transfer,[106] and the Framework for the Construction of Water Rights System.[107]

Moreover, the local people's congresses at the provincial level have formulated or issued local specialised and detailed regulations to implement the China Water Law 1988 and other laws and administrative regulations, *e.g.*, the Detailed Implementation Rules for Water Drawing Licence System of Xinjiang and the Water Resources Management Regulations of Shanxi Province. Some local people's governments have issued local rules to implement laws and regulations and departmental rules pertinent to water. In total, more than 700 pieces of local specialised and detailed regulations or local rules had been made before the end of November 2001.

After the adoption of the China Water Law 2002, the legislative bodies other than the NPC or its Standing Committee have been enacting, amending or repealing relevant laws to implement it.[108] On 31 October 2002, the Standing Committee of the People's Congress of Zhejiang Province adopted the Water

---

102 See Jing, Z. (2002), *supra* note 95.

103 The Implementation Rules on Water Industry Policy was promulgated by the MWR on 11 June 1999, and came into force on the same day, its official Chinese version in *Gazette of the State Council of the PRC* (1999) 26, 1132-1142.

104 The Procedure Provisions on Inspection & Approval of Application for Water-Drawing Licence was promulgated by the MWR on 9 June 1994, and came into force on the same day, its Chinese version available at MWR webpage http://www.mwr.gov.cn/zcfg/Detail.wct?SelectID=6697&RecID=13 (last visited 31/05/2005).

105 The Guidelines on the Water Rights Transfer Concerning the Stem Stream of the Yellow River within Inner Mongolia and Ningxia was promulgated by the MWR on 18 May 2004, its Chinese version available at MWR webpage http://www.mwr.gov.cn/index/20040521/34115.asp (last visited 31/05/2005).

106 The Opinions on Water Rights Transfer was promulgated by the MWR on 11 January 2005, its Chinese version available at http://www.mwr.gov.cn/index/20050117/45753.asp (last visited 31/05/2005).

107 The Framework for the Construction of Water Rights System was promulgated by the MWR on 11 January 2005, its Chinese version available at MWR webpage http://www.mwr.gov.cn/index/20050117/45755.asp (last visited 31/05/2005).

108 See Wang, S. (2004) *Insisting Scientific Development Concept, Comprehensively Promoting and Implementing Sustainable Development Water Conservation*, 7 January 2004, available at MRW webpage http://www.mwr.gov.cn/bzzs/20040107/29102.asp (last visited 31/05/2005).

Resources Administration Regulation of Zhejiang Province at its 39th Session.[109] This was the first local regulations under the China Water Law 2002 to be implemented at the provincial level.

Under the provisions of current Chinese water law, the three aspects of water rights are recognised, explicitly or implicitly, and are to some extent integrated, although certain issues remain outstanding and require further attention.

3.3 DEFINITION OF WATER RESOURCE

By adopting an analytical approach to the legal provisions, this section will identify issues concerning the definition of water resource. This is a necessary foundation for the subsequent comparison of the relevant provisions relating to the definitions of water resource in international law and/or the domestic law of selected jurisdictions.

3.3.1 What is water resource?

Article 2 (2) of the China Water Law 2002 provides that "[t]he water resource referred to in this Law includes surface water and groundwater".[110] This definition, however, is insufficient, since no provisions have been made to deal with the scope of surface water as well as groundwater.

Further, the general definition of water resource in this article is inconsistent with other provisions of the same law, as well as with the normative content of other legal documents. Firstly, for example, from Article 48 of the China Water Law 2002, one may conclude that water resource includes the waters in rivers, lakes or ground aquifers.[111] Secondly, Article 26 of the China Water Law 2002 provides that "[t]he State shall encourage the exploitation and/or utilisation of

---

109 The Water Resources Administration Regulation of Zhejiang Province came into force on 1 January 2003. Its Chinese version available at http://www.ep.net.cn/cgi-bin/dbfg/doc.cgi?id=1349 (last visited 31/05/2005).

110 See art. 2 (2), China Water Law 2002.

111 Article 48 of the China Water Law 2002 reads as follows:

> "Any unit or individual who draws water directly from rivers, lakes or ground aquifers shall, in accordance with the State provisions relating to the water-drawing licence system and the paid use system on water resources, apply to the relevant competent department of water administration or basin administrative authority for a water-drawing licence, pay the water resources fee regulated, and then obtain a water-drawing right. However, the drawing of water for a household or for livestock, a small quantity for poultry drinking or other small quantity water drawings, are exempted."

hydraulic power resource", while Article 27 requires that "[t]he State shall encourage the exploitation and/or utilisation of water transport resource". From the content of these two provisions, it seems that hydraulic power resources and water transport resources are also water resources. Thirdly, the MWR was prescribed by the State Council as its competent department of water administration in 1988. It has been further enforced since the new institutional reform of the State Council that was introduced in 1998.[112] In accordance with the stipulations of the State Council, one of the mandates of the MWR is the "[i]mplementation of integrated management of water resources, including atmospheric water, surface water and groundwater".[113] It is clear that atmospheric water is included in the scope of water resources, and this is also inconsistent with the scope of water resources regulated in the China Water Law 2002.

In brief, as demonstrated above, the problem is that the provisions regulating the definition of water resource in the China Water Law 2002 are inconsistent with each other, and are also incompatible with other relevant legal documents.

3.3.2 Difference between "water" and "water resource"?

Article 3 of the China Water Law 2002 reads as follows:

> "Water resources shall be owned by the State. The ownership of water resources shall be exercised by the State Council on behalf of the State. The water of a pond owned by or of a reservoir built and managed by a rural collective economic organisation shall be utilised by the organisation."[114]

The question is whether "water" and "water resource" are the same subject in law. Moreover, is there any difference between "water" and "water resource"? If no difference exists, why are both terms employed, and what is the legal status of bottled water? This problem will be examined further in the next section.

---

[112] See Regulation on Functions Allocated, Internal Institutions and Personnel of Ministry of Water Resources, Guoban [1998]87. It was approved by the State Council and promulgated on 23 June 1998 to implement the Decision on the Plan for the Institutional Reform of the State Council, adopted at the First Session of the Ninth NPC on 10 March 1998. In accordance with the above Regulations, the former Ministry of Geology and Mineral Resources handed over the administration of groundwater to the MWR.

[113] *Id.*, section 2(2). See also MWR, *Main Responsibilities* (English version), available at http://www.mwr.gov.cn/english/about_mwr/about_mwr_mr.htm (last visited 31/05/2005).

[114] See art. 3, China Water Law 2002.

3.4 PROPERTY RIGHT ON WATER RESOURCES

Using an analytical approach to the provisions of laws as well as practice including cases, this section will identify water rights problems with reference to property rights. This is fundamental to make a comparative analysis of the relevant provisions of the property right on water resources in international law and/or the domestic law of selected jurisdictions.

3.4.1 Ownership on water resources

3.4.1.1 Legal provisions

The legal system in China belongs to the family of civil law, as mentioned above. In theory, legal rights and obligations pertinent to property are and should be expressly stipulated by law. The definition of property ownership may be found in the General Principles of the Civil Law 1986.[115] Property ownership refers to the comprehensive right of the owner, which includes the right to lawfully possess, utilize, profit from and dispose of his property.[116]

China is a socialist country, and this characteristic is reflected in the concept of socialist public ownership as stipulated by law, *i.e.*, State ownership, namely ownership by the whole people, and collective ownership by the working people. These play an important, even vital, role in society. This also applies to water or water resources. In accordance with the China Water Law 2002, there is only one kind of ownership on water resources, *i.e.*, state ownership, which shall be exercised by the State Council on behalf of the State.[117]

3.4.1.2 Practice

Practice in the industry of agriculture[118]

---

[115] The General Principles of the Civil Law of the PRC, was adopted at adopted at the 4th Session of the Sixth NPC on 12 April 1986, and came into force on 1st January 1987. General Principles of the Civil Law of the PRC, official Chinese version, *Gazette of the Standing Committee of the NPC* (1986) 4, 3-25.

[116] *Id.*, art. 71. It provides that, " '[p]roperty ownership' means the owner's right to lawfully possess, utilise, profit from and dispose of his property".

[117] See art. 3, China Water Law 2002.

[118] In accordance with Article 2 of Agricultural Law of the People's Republic of China (1993), "agriculture" as mentioned in this section includes crop-plantation, forestry, animal husbandry and fishery. The Agriculture Law of the PRC was primarily adopted at the 2nd Session of the Standing Committee of the Eighth NPC on 2nd July 1993, and came into force on the same day. On 28 December 2002, it was amended at the 31st Session of the Standing Committee of the Ninth NPC, and the

For the purpose of ensuring the fundamental position of agriculture in the national economy, developing the socialist market economy in rural areas, and promoting the continuous, steady and coordinated growth of agriculture, the Chinese government has adopted and implemented numerous polices. One of the most important aspects is the contract operation and responsibility system, and relevant laws have been enacted to support these.[119] Rural land is subject to the contract operation system. In accordance with the Rural Land Contract Operation Law of the PRC, rural land refers to cultivated land, forestland, grassland and other lands employed in agriculture in accordance with the law, including waste mountains, waste digs, waste hills and waste beaches etc., owned by collectives or the State and lawfully exploited by agricultural economic organisations.[120]

In their production activities, individuals or collectives sometimes collect diffused water into the ponds or small dams that they have built or that have been built by others for the purpose of developing the contractual areas, within which those individuals or collectives are entitled to engage in agricultural activities. Those ponds or small dams are, however, located within the contractual areas.

The Yiwu-Dongyang water transaction case[121]

This case and the following one are related to the transfer of "water rights", which has not been recognised by law precisely. However, these two cases appeared. It is therefore necessary to discuss them, to find the legal issue(s) arising from them, and further to incorporate the relevant legal issues into a suggested water rights system.

---

amendment came in force on 1st March 2003. Agriculture Law of the PRC (Revision), official Chinese version, *Gazette of the Standing Committee of the NPC* (2003) 1, 20-30.

119 For example, in Chapter Two "System of Agricultural Production and Operation" of Agriculture Law of the PRC (1993) gave more provisions. On 29 August 2002, Rural Land Contractual Farming Law of the PRC was adopted, and came into force on 1st March 2003, and in which the contractual farming system has been adjusted in more detail. On 28 December 2002, Agricultural Law of the PRC was amended, and the Rural Land Contractual Farming Law of the PRC is regarded as a special law to the Agriculture Law of the PRC (Revision). Rural Land Contractual Farming Law of the PRC, official Chinese version, *Gazette of the Standing Committee of the NPC* (2002) 5, 347-352.

120 See arts. 2-3, Rural Land Contractual Farming Law of the PRC.

121 Although this case and Zhangze Water Transaction case happened before the adoption of the China Water Law 2002, there is value in discussing them. Firstly, whether in accordance with the China Water Law 2002 or the China Water Law 1988, the water resources concerned in them are both are state-owned. Secondly, the water-drawing licence system applied at that time is still valid.

Well known for its marketing of small-size merchandise, Yiwu City of Zhejiang Province, in the lower reaches of the Jinhua River, is so short of water that people often have to carry buckets of mineral water to their homes for drinking and cooking. On 24 November 2000, Yiwu City and Donyang City (which is located in the upper reaches of the river) in the same province signed an agreement for the paid transfer of the said water right pertinent to property.[122] After the construction of water works relating to the transaction were completed, Dongyang City has been delivering water to Yiwu City since 6 January 2005.[123] According to the agreement:

(1) For the purpose of annually buying the right to use 49.999 million m^3 of water in the Hengjin Reservoir located within the jurisdiction of Dongyang City, Yiwu City shall pay CNY 200 million in total.
(2) The transfer of the right to use the waters does not make any change in the ownership of the Reservoir, the operation and the maintenance of the Reservoir being the responsibility of Donyang City.
(3) Yiwu City shall pay a comprehensive charge (in which the water resources charge is included) for the waters actually supplied per year at the standard of CNY 0.10 / m^3.
(4) The water works diverting the waters from Hengjin Reservoir to Yiwu City have to be planned, designed, invested and constructed by Yiwu City. However, the policy affairs and construction works pertinent to the water works within the jurisdiction of Dongyang City are for Dongyang City to deal with, the relevant cost being paid by Donyang City.[124]

The above transaction has led to a vehement argument in China, in which, however, it appears that no or very few experts in law participate. In general, there are two opinions. Mr. Wu, Guoping may be taken as a representative of the first argument. He asserts that, (a) the transfer of water rights pertinent to property between the two cities is not in accordance with the present law as it contravenes the system of the state-ownership on water resources, and (b) the paid transfer of water rights pertinent to property between different administrative regions will lead to a series of legal problems.[125]

---

[122] See Shen, H. and Wang, L. (2001) The First transaction on water-use rights was concluded in China, *People's Daily*, 16 February 2001.

[123] Pang, H. and Zhu, Y. (2005) The First National Water Rights Transaction Began to Deliver Goods, Chinese version, available at http://www.xj.xinhuanet.com/2005-01/06/content_3519359.htm (last visited 31/05/2005).

[124] See Wang, L. (2001) The first transaction on water rights was concluded in Zhejiang Province, *China Water Resources News*, 10 February 2001.

[125] See Wu, G. (2001) A Letter to the Minister Wang Shucheng, *China Water Resources News*, 28 February 2001.

On the contrary, Wang, Shucheng, the Minister of Water Resources, on the opposing side of the debate, considers that, innovations should not be excluded automatically simply because there is no existing legal basis for them, and new laws and regulations should be enacted to govern these.[126] However, as discussed in Chapter 1, there is no provision concerning the transfer of water rights in the China Water Law 2002, which was adopted at the 29th Meeting of the Standing Committee of the Ninth NPC on 29 August 2002.

The Zhangzhe water transaction case

In June 2001, with the coordination of the Upstream Management Bureau of the Zhanghe River of the Haihe River Water Conservancy Committee, the three provinces of Henan, Hebei and Shanxi concluded an agreement on inter-provincial water buying. Anyang City of Henan Province and Handan City of Hebei Province bought 20 million m^3 of water from Shanxi Province, which solved the urgent problem of a shortage of farming water along the riverbanks in the former two provinces. It was reported that a good method was found for dealing with the surplus water in the Zhangzhe Reservoir in Changzhi County of Shanxi Province resulting from flooding, and this is a successful case of achieving the inter-regional allocation of water resources through economic means, being an effective supplement to, or even a beneficial breakthrough of, administrative allocation. Such practices and explorations serve as a forerunner to the use of the market mechanism for transferring water rights pertinent to property and signal that paid transfer may become a focus of water rights pertinent to property transfer in at some time in the future.[127]

From then on, similar water transactions were implemented in March 2002, and from March to June 2003.[128]

3.4.2 Right to draw water and use right on water resources

At present, strictly speaking, there is no provision in law to regulate the right to draw water or the use right on water resources, as a type of property right, although a water-drawing licence system has been implemented. However, the right to draw water has been discussed as a type of property right academically, and has also been expected to be provided for in a proposed Property Law of the

---

126 See Wang, S. (2001) A Reply to Wu Guoping, *China Water Resources News*, 28 February 2001.

127 This paragraph is summarised from the news report "Trans-provincial paid water diversion succeeded in Zhanghe River", written by Mong, Z. and Hu, F (2001), in *China Water Resources News*, 18 May 2001.

128 See Ma, J.(2002) Zhanghe upstream bureau managed the organization of inter-provincial water transaction again, *China Water Resources News*, 19 March 2002; see also Niu, F. and Hu, F.(2003) Positively exploring new routes to resolve the water disputes in Zhanhe Basin, *China Water Resources News*, 22 May 2003.

PRC, on the one hand.[129] On the other, the MWR has organised some regions to implement the transfer of water rights under the name of developing a water-saving society, and has also formulated some specific documents to guide the transfer. Therefore, the water-drawing licence system in China as well the relevant documents issued by the MWR will be introduced, and the nature of the right to draw water will be discussed.

3.4.2.1 Right to draw water and water-drawing licence system

China has established a system to govern the activity of drawing water. According to Chapter IV "Management of Water Use" of the China Water Law 1988, China should implement a water-drawing licence system. However the State Council has the power to decide the steps to be taken, the scope of and the implementation measures of the water-drawing licence system.[130] On 1st August 1993, five years after the adoption of the China Water Law 1988, the Measures on Implementation of Water-Drawing Licence System, a piece of administrative regulations, was promulgated by the State Council. Nationwide, the MWR has enacted several departmental rules, in which detailed methods and procedures are provided, to implement such a water-drawing licence system, *e.g.*, Procedure Provisions on Inspection & Approval of Application for Water-Drawing Licence (9 June 1994), Provisions on Water Quality Management of Water-Drawing Licence (December 1995),[131] and Provisions

---

[129] See arts. 972-973 in Section Eight "Royalty Usufractuary", Book Five "Property Law" of Chinese Civil Code (Suggestion Draft). Wang, L., ed. (2004) *Chinese Civil Code (Draft): Suggestion and Explanation*, China Legal System Press, Beijing. Article 972 suggests that, "[W]ater-drawing right refers to the right which entitles a citizen, legal person or unit, with the approval in accordance with law, to withdraw water directly from rivers, lakes or ground aquifers with utilising water projects or mechanical facilities".

[130] See art. 32, China Water Law 1988. One should note that, the China Water Law 2002 has a similar provision, *i.e.*, Article 7. It reads as follows:

> "The State shall implement a water-drawing licence system and a paid use system on water resources in accordance with the law, with the exception of the water of ponds or reservoirs belonging to rural collective economic organisations utilised by these organisations or their members. The competent department of water administration under the State Council shall be in charge of the organisation and implementation of these two systems in the country."

[131] The Provisions on Water Quality Management of Water-Drawing Licence, its Chinese version available at MWR webpage http://www.mwr.gov.cn/zcfg/Detail.wct?SelectID=6697&RecID=5 (last visited 31/05/2005).

on Supervision & Management of Water-Drawing Licence (29 July 1996).[132] Although the China Water Law 2002 was adopted on 29 August 2002, these regulations and rules still remain valid and active, because they have not been abolished.

Water-drawing licence system: object

Under the water-drawing licence system, "water-drawing" means drawing water directly from rivers, lakes or ground aquifers with the aid of water installations or machinery.[133] However, not all water-drawing is subject to the water-drawing licence system; there are exceptions and exemptions. Exceptions include: (a) household use, livestock and poultry drinking; (b) small quantity water-drawings for agricultural irrigation; and (c) small quantity water-drawings by the power of human beings, livestock and other methods.[134] As to exemptions, they include: (a) where water-drawing is necessary for agricultural drought-relief in case of urgency; (b) where water-drawing is necessary to safeguard the security of the construction and the production of underground projects, *e.g.*, mineral wells; and (c) where water-drawing is necessary to defend or eliminate hazards to the public security or the public interest.[135]

Water-drawing licence system: principle

Because the volume of water is not always sufficient for simultaneous uses of different kinds, and conflicts are unavoidable, it is necessary to regulate the relationships between them, *i.e.*, what kind of uses should be given priority, and this is the principle of the water-drawing licence system. The author believes that these may be deduced from the provision of Article 21 of the China Water Law 2002.[136] That is to say, the general priority ranking principle shall be (a) water-use demands for the livelihood of urban and rural inhabitants, (b) agricultural, industrial and eco-environment water-use demands, and (c) navigation and other demands.

Water-drawing licence system: management

In order to implement the water-drawing licence system, the following are provided for the administration of the system.

(1) Plan-making. The National Development and Reform Commission (NDRC) and MWR are in charge of the national macro allocation of water resources. MWR formulates national and trans-provincial

---

132 The Provisions on Supervision & Management of Water-Drawing Licence, its Chinese version available at MWR webpage http://www.mwr.gov.cn/zcfg/Detail.wct?SelectID=6697&RecID=2 (last visited 31/05/2005).

133 See art. 2, Measures on Implementation of Water-Drawing Licence System.

134 *Id.*, art. 3.

135 *Id.*, art. 4.

136 See art. 21, China Water Law 2002.

medium-term and long-term water supply-demand plans in conjunction with other relevant departments under the State Council. These plans shall be submitted to NDRC for examination and approval, and then be executed. The competent department of water administration (CDWA) of the local people's government at or above the county level, in conjunction with other relevant departments at the corresponding level, formulate its local medium-term and long-term water supply-demand plans, in accordance with the medium-term and long-term water supply-demand plans at the next higher level and the regional practice, these plans shall be submitted to the competent department of development and planning administration at the corresponding level, and then be executed.[137]

(2) Authorities. MWR and its authorised basin administrative authorities, and the CDWAs of the local people's governments and its authorised departments.[138]

(3) Procedure. Where it is necessary to apply for a water-drawing licence for a new project, an extension project, or a modification project, the construction unit must make a pre-application for a water-drawing licence before submitting the feasibility study report to the government for approval.[139] After the project has obtained the approval from the government, the construction unit may apply for a water-drawing licence through presenting required documents.[140] The authority, which accepts the application form and other documents required, shall inspect the application, and make a decision either to issue or not issue a water-drawing licence in accordance with the law.[141] When a decision is taken not to issue a water-drawing licence, the applicant may apply for reconsideration to the next higher authority.[142] However, if the decision not to issue a water-drawing licence is upheld, the applicant my initiate

---

137 *Id.*, art. 44.

138 See arts. 10-11, 15 and 19, Measures on Implementation of Water-Drawing Licence System; also art. 3, Procedure Provisions on Inspection and Approval of Application for Water-Drawing Licence.

139 See art. 5, Procedure Provisions on Inspection and Approval of Application for Water-Drawing Licence.

140 *Id.*, arts. 10-11.

141 *Id.*, arts. 12-17.

142 *Id.*, art. 18.

a legal action against the authority.[143] The valid period for a water-drawing licence must not exceed five years.[144]

(4) Supervision. In accordance with the stipulations of the Provisions on Supervision & Management of Water-Drawing Licence, the competent organs are under a duty to supervise the implementation of the water-drawing licence system.[145] Supervision should be carried out in relation to the facilities employed by the applicants, the quantity of water drawn by the applicants and the quality of the water drawn.[146] If the applicant performs any illegal activity, the competent organs are under a responsibility to carry out an investigation into legal liability in accordance with the law.[147]

3.4.2.2 MWR's actions relating to water rights and its transfer

Chinese politics often appears puzzling, as may be illustrated by the MWR's practice relating to water rights. As mentioned earlier, the recommendation that transfers of water-drawing rights as contained in the Water Law of the PRC (Amendment Draft) is not encompassed in the China Water Law 2002, which was adopted on 29th August 2002.[148] This has not, however, prevented the MWR from implementing its ideas or policies concerning water rights transfer. These may be discussed through its relevant practice and documents.

Water rights and its transfer practice in test regions

Since July 2001, the MWR has practiced its water rights transfer policy under the name of "developing water-saving society" or "water-saving and pollution-preventing society."[149] Initially, it selected Zhanhye City (in Gansu Province)

---

143 *Id.*

144 *Id.*, art. 23.

145 See art. 4, Provisions on Supervision & Management of Water-Drawing Licence.

146 *Id.*, art. 26.

147 *Id.*, arts. 41-42.

148 See the relevant discussion in Chapter 1.

149 Water-saving is considered as one of important principles in China Water Law 2002, and this is reflected in Chapter One "General Provisions" (see arts. 4, 8, 10-11 and 13), Chapter Two "Planning for Water Resources" (see art. 14), Chapter Five "Allocation and Saving Utilisation of Water Resources" (see arts. 50, 52-53) and Chapter Seven "Legal Liability" (see art. 71). In particular, Article 8 reads as follows:

> "In order to establish a water-saving society, the State shall strictly enforce water-saving activities, greatly promote water-saving measures, spread the application of new water-saving technology and/or techniques, and develop water-saving industries, agriculture and/or service industries.

as a test region in July, and water rights transfers were implemented in several large irrigation areas in this city.[150] In December 2002, the MWR and the People's Government of Gansu Province officially approved the Zhangye City as a national test region for developing a water-saving society, and subsequently, Mianyang City (in Sichuan Province) and Dalian City (in Liaoning Province) were approved by the MWR and the corresponding provincial people's governments as national test regions for developing a water-saving society chronologically.[151] According to a notice issued by the MWR, up to 30 June 2004, there were also 91 provincial test regions for developing a water-saving society.[152] However, the term "water rights" is inconsistently employed in different regions or written materials. In October 2003, for example, Shucheng Wang, the Minster of MWR, spoke about water use right and its transaction, but neither the legal form nor substantial content of water use right were given.[153] As discussed in Chapter 2, the term "water use" is an over-used expression with many different meanings.[154] Further, an excessive distinction between water use right and the ownership of water is inappropriate in the present situation.[155]

According to the practice in the Liyuanhe Irrigation Area of the Zhangye test region, the water rights system employed may be summarized as follow. Based on the size of the irrigation land, population and household animals, applying the water use quota for different water users or industries, the available water is allocated to every water user. The 10,680 water users are organised into 275

> The people's government at various levels shall adopt measures to strengthen the management of water-saving activities, establish a development and promotion system for water-saving technologies, and foster and develop water-saving industries.
>
> Every unit or individual is under an obligation to undertake water-saving activities."

150 See Wang, S. (2003) Working Key-points in Developing Water-saving Society, in *Collection of Materials on National Test Regions for Developing Water-saving Society*, China WaterPower Press, Beijing, 2004, p. 3.

151 See Water Resources Bureau of the MWR and National Water-saving Bureau (2004) *Collection of Materials on National Test Regions for Developing Water-saving Society*, China WaterPower Press, Beijing, p. 1.

152 See MWR, *Notice on Publication of Provincial Test Regions for Developing Water-Saving Society*, its Chinese version available at the MWR webpage http://www.mwr.gov.cn/index/20040630/38568.asp (last visited 31/05/2005).

153 See Wang, S. (2003), *supra* note 150, pp. 3-5.

154 See the detailed discussion in 2.1.2.1 of Chapter 2.

155 *Id.*

water-user groups, which are further classified into 45 water-use associations. The government has issued a certificate for water resources use rights to each water user. Such a certificates records the name of the holder, amount(s) of item(s)/person(s) that need water, the corresponding water use quota, the annual quantity of water entitlement, etc. The holder of a certificate is entitled to buy water ticket(s) each year, which allows him not only to actually draw and use water, but is also negotiable under the regulated management. Moreover, regarding transaction quantity, a transaction of up to 1,000 m^3 shall be concluded through the relevant water-user-association, a transaction from 1,001 m^3 to 10,000 m^3 shall be concluded through the relevant water management station, a transaction from 10,001 m^3 to 100,000 m^3 shall be concluded through the Liyuanhe Water Management Office, and a transaction over 100,000 m^3 shall be concluded through the relevant competent department of water administration at or above the county level. As to the transaction price, this is for the market to decide and for the transaction parties to negotiate. However, the government implements macro-control, i.e., the price per m^3 for a transaction must not be more than three times the original price per m^3.[156]

Water rights and its transfer regulated in MWR's documents

There are three documents issued by the MWR directly concerning water rights transfer, namely, the Guidelines on the Water Rights Transfer Concerning the Mainstream of the Yellow River within Inner Mongolia and Ningxia, the Opinions on Water Rights Transfer, and the Framework for the Construction of Water Rights System. However, the term "water rights" are incompatibly employed, even in the same document. In the first document, the term "water rights" refers to the right to draw water. After a unit or individual who draws water resources directly from rivers, lakes or underground obtains a water-drawing licence and pays water resources fee, he then obtains the right to draw water. When a holder of the right to draw water saves water through engineering measure(s), the holder is entitled to transfer the right to draw the saved water.[157]

Under the Opinions on Water Rights Transfer, "water rights" means the right to use water resources,[158] but the Framework for the Construction of Water Rights System issued on the same day offers different and confused meanings. Although the latter does not expressly define the term "water rights", one may

---

[156] Liyuanhe Water Management Office (2003) Making Water Rights Clear, Allocating Water to Each Water User, in *Collection of Materials on National Test Regions for Developing Water-saving Society*, China WaterPower Press, Beijing, 2004, pp. 49-55.

[157] See paras. 1 and 3 of Part Two, Guidelines on the Water Rights Transfer Concerning the Mainstream of the Yellow River within Inner Mongolia and Ningxia.

[158] See the Preface, Opinions on Water Rights Transfer.

conclude that is includes the ownership of water resources and the use right on water resources.[159] It also provides, however, that water rights transfer refers to the transfer of the use right on water resources.[160]

3.4.3 Problems identified

3.4.3.1 Ownership other than the State ownership

As stated above, in the activities of their production under contracted responsibility, individuals or collective organisations sometimes collect diffused water into the ponds or small dams that they have built, or those have been built by others, for the purpose of developing the contractual areas, within which these ponds or small dams are located. The question is: who is the owner of the waters in such ponds or small dams? Is it the State or the contracting parties other than the governments or its authorised units concluding the contracts? If the State is the owner, then what right does the latter have? Additionally, when a private person buys a bottle of water, who owns the bottled water? Is it the State, or is it the private person? In brief, the problem is that there is, in fact, individual or collective ownership on water or water resource, although this is not recognised by/in law.

3.4.3.2 Waters in reservoirs

Considering the practice in agriculture, another problem arising from the provision of the Article 3 of the China Water Law 2002 is how to distinguish the waters in a reservoir owned by a rural collective economic organisation from the waters of the river in which the reservoir is located. Further, is such a reservoir built within the lands owned by a rural collective economic organisation different from that built within the lands owned by the State? Is such a reservoir built upstream different from that built downstream or midstream?

3.4.3.3 Holder of the State ownership

In the Yiwu-Dongyang and the Zhangzhe water transaction cases mentioned above, the transferred water or water resources were all state-owned in accordance with the law. The problem is that, if this is so, why are payments

---

159 See para. 2 of Part Two, Framework for the Construction of Water Rights System. It provides that "water rights system consists of the ownership system on water resources, the use right system on water resources and the transfer system on water rights".

160 See para. 1 of Sub-part Three of Part Two, Framework for the Construction of Water Rights System.

being made between different local governments, as this implies that some form of ownership has been passed between these local governments? In other words, under state ownership, does a precise owner exist?

3.4.3.4 Transfer of water property rights

In the above two cases, what was transferred in fact? In other words, for what was the buyer paying? Was it the waters in the water works or was it the use of the water works? Alternatively, was it the service supplied by the owners or operators of the water works? It should be noted that, in the Zhangzhe water transaction case, the Zhangzhe Reservoir was designed and constructed for flood control, so that even if no one buys the said water right pertinent to the property, the water in that reservoir has to be abandoned before the beginning of the flood season. In brief, the problem is to determine what kinds of water rights pertinent to property may be transferred.

3.4.3.5 Nature of the right to draw water

In academic discussions, some Chinese legal scholars of civil law consider that the right to draw water under a licence belongs to usufruct.[161] However, this is a misunderstanding of the conception for the following two reasons.

Firstly, usufruct is "[t]he right of enjoying a thing, the property of which is vested in another, and to draw from the same all the profit, utility and advantage which it may produce, provided it be without altering the substance of the thing. The obligation of not altering the substance of the thing, however, takes place only in the case of a complete usufruct".[162] Usufructs are of two kinds, *i.e.*, perfect and imperfect. "Perfect usufruct, which is of things which the usufructuary can enjoy without altering their substance, though their substance may be diminished or deteriorated naturally by time or by the use to which they are applied; as a house, a piece of land, animals, furniture and other movable effects. Imperfect or quasi usufruct, which is of things that would be useless to the usufructuary if be did not consume and expend them, or change the substance of them, as money, grain, liquors."[163] The question is, if the right to draw water is usufruct, then what is the relevant property, the licence or the water (resources)? It is obvious that the licence itself could not be the property to be usufructed, but if the property is water (resources), when the utilisation is consumptive, especially where water is to be sold or exported, the holder of the

---

[161] See arts. 380-381 in Section Seven "Royalty Usufractuary" of China Property Law (Suggestion Draft). Wang, L., ed. (2001) *supra* note 129.

[162] See *Bouviers Law Dictionary* (1856 ed.), available at http://www.jusbelli.com/Bouvier/bouvier1856_u_all.html (last visited 31/05/2005).

[163] *Id.*

licence can neither restore the water drawn nor return the same quantity of water in the same quality.

Secondly, the right to draw water under a licence belongs to administrative law by its very nature, rather than civil law, under a civil law system.[164]

3.4.3.6 Water rights: an inconsistent term from MWR

From both the practice on water rights and its transfer in the test regions organized by the MWR, and the MWR's documents that regulate water rights and its transfer, the term "water rights" is employed inconsistently. If the MWR does not understand the meaning(s) of "water rights", but regulates them nonetheless, how can it be concluded that this water rights system is appropriate for implementing and helpful for promoting sustainable integrated water resources management.

3.5 THE HUMAN RIGHT TO WATER

Employing an analytical approach to the provisions of laws as well as the relevant practice, this section will identify water rights problems relating to the human right to water, for this is fundamental for comparing the relevant provisions of the human right to water in international law and/or the domestic laws of selected jurisdictions.

3.5.1 Legal provisions

In this section, the provisions relating to the human right to water in Chinese water law will be discussed, and this is a foundation for identifying problems in Chinese water law, as well as comparing the relevant provisions in international law and/or the water laws of selected jurisdictions. After the general legal provisions relating to the human right to water have been discussed, the drinking-water sources conservation system will be examined.

3.5.1.1 General provisions

As discussed in Chapter 2, access to water to meet basic human needs is an important component of human rights. Although the human right to water has

---

[164] See the opinions of experts from Germany, Italy in Sino-Germany International Forum on Chinese Property Law (Suggestion Draft), available at available at Chinese Civil and Commercial Law Web http://www.civillaw.com.cn (last visited 31/05/2005).

not been recognised expressly in China, there are many laws and policies relating to water that emphasise the supply of water for domestic use.

On 13 August 1984, the State Council approved the Report on Accelerating to Resolve the Problem of the Domestic and Livestock Drinking Water in Rural Areas, and the Temporary Rules on Work concerning Domestic and Livestock Drinking Water in Rural Areas.[165] Both of these were drafted by the Ministry of Water Resources and Electricity. Under the China Water Law 1988 as well as the China Water Law 2002, in the development and utilisation of water resources, the domestic water use of urban and rural inhabitants shall be satisfied first,[166] and drawing water for domestic water use fell into the exceptions of the water-drawing license system.[167] According to Article 4 of the Water Industry Policy, domestic and livestock drinking water is one of the important aspects in the development of the water industry during its implementation.[168] In the Implementation Rules on Water Industry Policy, the shortage of water in rural and urban areas being effectively lessened and the difficulty of domestic and livestock drinking water in rural areas being basically resolved are listed among the major targets for the development of water resources in 2010.[169]

According to Article 4 of the China Water Law 2002, livelihood water-use, production and business water-use, and ecological-environmental water-use should be coordinated. Further, the China Water Law 2002 provides that, in the development and utilisation of water resources, the domestic water demands of urban and rural inhabitants shall be satisfied first.[170] Drawing a small amount of water for household use is excluded from the water-drawing licence system and the water resource paid use system.[171] Under the Grassland Law of the PRC, people's governments at or above the county level should encourage the construction of pasture water conservation facilities, develop pasture water-saving irrigation, and enhance the condition of drinking-water.[172]

---

165 The Report on Accelerating to Resolve the Problem of the Domestic, Livestock Drinking Water in Rural Areas, and the Temporary Rules on Work concerning Domestic and Livestock Drinking Water in Rural Areas, their Chinese version available at http://www.sxwater.gov.cn/zcfg/NCJSYSFLFG/NCJSYSFLFG1015.htm (last visited 31/05/2005).

166 See art. 14, China Water Law 1988; art. 21, China Water Law 2002.

167 See art. 32 (1), China Water Law 1988; art. 48 (1), China Water Law 2002.

168 See art. 4, Water Industry Policy.

169 See art. 4 (2), Implementation Rules on Water Industry Policy.

170 See art. 21, China Water Law 2002.

171 *Id.*, art. 48.

172 See art. 28 (2), Grassland Law of the PRC (Revision).

To conclude these above provisions, it is clear that, in Chinese water law, (a) domestic water use has the first priority among all water uses, and (b) resolving the difficulty of drinking water in rural areas is emphasised. That is to say, the human right to water, in particular for the poor in rural areas, is recognised.

3.5.1.2 Drinking-water sources conservation system

Needless to say, drinking water is one of the key elements of the human right to water. In order to protect the sources of drinking water, a drinking-water sources conservation system is provided by the China Water Law 2002. The main content of the system is: it is the duty of the people's governments at the provincial level to delimit drinking-water sources conservation zones, to take measures to prevent water sources from depletion and pollution, and to guarantee the safety of drinking-water for urban and rural inhabitants.[173]

Under the Law for Water Pollution Control of the PRC, protection zones of all grades (logically, drinking-water sources conservation zones shall be included) shall be indicated by clear geographic demarcations.[174] Further, in accordance with the China Water Law 2002, no polluted water drainage sources shall be located in the drinking-water sources conservation zones.[175] In case of emergency, such as the severe pollution of a drinking-water source which threatens the safe supply of water, the environmental protection department shall, with the approval of the people's government at the same level, take compulsory emergency measures, including ordering the enterprises or institutions concerned to reduce or stop the discharge of pollutants.[176]

As to legal liability, Article 67(1) provides that the local people's government at or above the county level shall order the dismantling of any pollutant outlets built in a drinking-water source conservation zone and shall order the restoration of the original condition within a prescribed period. Failing that, they shall be dismantled and the original condition restored forcefully. A fine of between 50,000 and 100,000 CNY shall be imposed. It is clear that, the human right to water is protected by Chinese water law.

---

[173] See art. 33, China Water Law 2002. It should be noted that, no relevant provisions were adopted under the China Water Law 1988 and Law for Water Pollution Control of the PRC. Even under Law for Water Pollution Control of the PRC (Revision), Article 20 (1) only provides that "The people's governments at or above the provincial level may delineate surface sources protection zones for domestic and drinking water according to law".

[174] See art. 20 (1), Law for Water Pollution Control of the PRC (Revision).

[175] See art. 34, China Water Law 2002; also art. 27 (1), Law for Water Pollution Control of the PRC (Revision).

[176] See art. 21, Law for Water Pollution Control of the PRC (Revision).

3.5.2 Practice

An official Chinese document shows that, in some areas of China, a severe shortage of water existed in history, especially in the years of drought, and these areas are distributed mainly in the mountains, the Yellow Soil Plateau, the coastal areas and the grass areas. Since 1st October 1949, the Chinese government has adopted several measures to resolve this problem. Before the end of 1979, the shortage of drinking water for 40 million persons had been resolved. From 1980 to 1983, the shortage of drinking water for 26,550,000 persons was resolved. However, at the end of 1983, there still remained over 56 million persons who were short of drinking water in every province level region, with the exception of Shanghai City. In some stony areas of the mountains and the Yellow Soil Plateau, it is more difficult to resolve the problem, and sometimes it was only temporarily resolved. Water pollution has also led to an increase in the number of persons who are short of drinking water.[177]

According to the Temporary Rules on Work concerning Domestic and Livestock Drinking Water in Rural Areas, the definition of the shortage of drinking water with reference to human beings may be summarised as: (a) in terms of distance, the nearest water-drawing point for a village is more than 2-4km from the village, or more than 100 meters at altitude; (b) in terms of quantity, during period of drought, the per capita per day water supply is under 10 kg in the north, and 40 kg in the south. As to the definition of the shortage of drinking water with reference to livestock, the quantity for larger animals is less than 20-50 kg per day, and for smaller ones is less than 5-20 kg per day.[178]

In accordance with the above standard, until the end of 1999, there still remained 24.23 million persons in rural areas who were short of drinking water, and therefore, a target was set by the Chinese government to resolve this problem during the period from 2000 to 2002.[179] The measures adopted include:

(1) Increasing investment. During the period, a CNY 5.7 billion fund originating from State Bonds was employed in the relevant project.
(2) Certain policies, *e.g.*, Management Rules and Implementation Opinions on Construction of the Rural Drinking Water Project, Technical Points

---

177 See Report on Accelerating to Resolve the Problem of the Domestic, Livestock Drinking Water in Rural Areas.

178 See arts. 2-3, Temporary Rules on Work concerning Domestic and Livestock Drinking Water in Rural Areas. Please note the water in quantity measured here is in kg because of the provisions of this rules.

179 See Bureau for Irrigation, Drainage and Rural Water Supply of the MWR (2003) *Historical Achievement in Rural Drinking Water Problem*, at http://www.chinawater.com.cn/ztgz/rcys/2/20030103/200301030113.asp (last visited 31/05/2005).

concerning Rural Drinking Water, were enacted, and some methods were implemented.

(3) Strengthening the management and supervision of the drinking water project. 580 thousand drinking water projects of different types in rural areas, including 100 thousand centralised water supply projects and 480 thousand small projects for single houses, were constructed. With these projects, the daily water supply could be increased by 2.4 million m^3, and therefore the majority of the people who had been listed as suffering from a persistent shortage of drinking water were able to access drinking water.[180]

However, many people were newly listed as short of drinking water because of severe drought in recent years, water pollution and other reasons.[181] At present, 21.65 million persons in rural areas of China are short of drinking water,[182] and over 300 million people, mainly in rural areas, can not access safe drinking water.[183]

3.5.3 Problems identified

Definite declaration of the human right to water

As mentioned above, due to serious water shortage, China has thousands and thousands of people without a safe, minimum water supply to meet their basic human needs every year because of severe drought, serious pollution and so on. The problem is that, although the Chinese government has adopted certain policies and implemented relevant measures to deal with this issue, the human right to water has not been definitely declared as a right, and is sometimes ignored or relegated in importance because of its weak status in law.

The first priority of the human right to water

Article 21(1) of the China Water Law 2002 provides that, "[i]n the exploitation and/or utilisation of water resources, the water-use demands for the livelihood of urban and rural inhabitants shall be satisfied first, while the

---

180 *Id.*

181 See Bureau for Irrigation, Drainage and Rural Water Supply of the MWR (2003) *Continuing to resolve the problem of rural shortage of drinking water*, available at http://www.chinawater.com.cn/ztgz/rcys/2/20030103/200301030113.asp (last visited 31/05/2005).

182 *Id.*

183 See Zhai, H. (2003) *struggling for a new situation on rural drinking water*, 5 January 2003, available at http://www.chinawater.com.cn/ztgz/rcys/1/20030108/200301080066.asp (last visited 31/05/2005).

agricultural, industrial and eco-environment water-use demands, as well as navigation demands etc., shall also be taken into consideration". One may conclude that the human right to water is given the first priority among all water uses. However, the problem is that, this article is in Chapter Three "Exploitation and Utilisation of Water Resources" of the China Water Law 2002, but not listed in Chapter One "General Provisions" as a principle. That is to say, the legal status of the human right to water is insufficient strong.

The content of the human right to water

Without the content of a right being provided, the holder of the right may not know what the right is, and others may not know how to avoid interrupting such a right. In China, the problem is simply that, there is no provision in law to clearly regulate the detailed content of the human right to water, which will be examined more in Chapter 4.

3.6 THE ENVIRONMENTAL RIGHT TO WATER

Using an analytical approach to the provisions of law as well as the practice including cases, this section will identify water rights problems relating to the environmental right to water in China. This is fundamental for comparing the relevant provisions of the environmental right to water, in international law and/or the domestic laws of selected jurisdictions.

3.6.1 Legal provisions

This section will consider the legal provisions relating to the environmental right to water in Chinese water law with respect to the general and special provisions. This knowledge is fundamental for identifying the problems concerning the environmental right to water in China.

3.6.1.1 General provisions

As mentioned in Chapter 2, the realisation of the environmental functions of water resources is very important.[184] To some extent, Chinese law has adopted some provisions to address the environmental water demand. Under the China Water Law 2002, for example, ecological-environmental water-use is among the three catalogues of water-uses recognised by law.[185] Further, the protection of water resources is linked with the ecosystem and/or environment.[186]

---

184 See the relevant discussion in 2.5 of Chapter 2.

185 The three catalogues of water uses recognised in the China Water Law 2002 include livelihood water-use, ecological-environmental water-use, and production and business water-use. See art. 4, China Water Law 2002.

186 *Id.*, art. 9. It reads as follows:

3.6.1.2 *Special provisions*

This section will discuss the special provisions in the China Water Law 2002 relating to the environmental right to water.

Firstly, the environmental right to water shall be reflected in the activities of formulating plans. The making of comprehensive plans for river basins, regional comprehensive plans, and special plans, which are of relevance to the utilisation of land, shall be coordinated with the relevant environmental protection plan and other plans, such as the development plans for national economy and society, the general plans for land use, and the urban general plans, and shall also take into consideration the needs in various regions and trades.[187] Regarding the formulation of the plans for the development and utilisation of water resources and those for the management of water resources, the departments of water administration, basin, administrative authorities and other relevant departments of the people's governments at or above the county level shall pay attention to maintaining the rational river flow and the rational water level of lakes, reservoirs and groundwater and to maintaining the natural purification capacity of the water system.[188] Further, the mid-term and long-term water supply-demand plans must be formulated on the basis of the current water supply-demand situation, corresponding national social and economic development planning, basin plan(s) and region plan(s), and be in accordance with the principles of coordinated water supply and demand, comprehensive equilibrium, ecological protection, strict saving and rational tapping of water sources.[189]

Secondly, the environmental right to water shall be taken into consideration in the development and utilisation of water resources. In detail, the ecological-environmental water-use must be taken into consideration, and full consideration shall be given to the ecological-environmental water-use need in the development and utilisation of water resources in arid and semi-arid areas.[190] Where an inter-basin diversion project is to be built, an overall plan and scientific justification must be provided, and consideration given to the demand for water in the basin which supplies the water and in the basin which receives

---

"For the purpose of protecting water resources, the State shall adopt effective measures to preserve natural flora, to plant trees and/or grow grass, to conserve water sources, to prevent and/or control soil erosion and water-body pollution, and to improve the eco-environment."

187 *Id.*, art.15 (2).

188 *Id.*, art. 30.

189 *Id.*, art. 44 (2).

190 *Id.*, art. 21.

it, while damage to the ecological environment shall be avoided.[191] In the development of hydropower stations, the ecological environment shall be protected, and the needs for flood control, water supply, irrigation, bamboo and log rafting, fishery, etc. shall be taken into account.[192] When permanent dams and sluice-gates are built on the migration routes of aquatic organisms, or on rivers which are navigable or suitable for bamboo and log rafting, the construction unit must at the same time build facilities for the passage of fish, ships and for bamboo and log rafting or, after approval by a department authorised by the State Council, take other remedial measures. The construction unit must also make adequate arrangements for the protection of aquatic organisms and for navigation and bamboo and log rafting during the construction period and the initial water-filling period, and also bear the expenses incurred thereby.[193]

Finally, to fulfil the environmental right to water is the duty of the government. The local people's governments at all levels shall strengthen their leadership of irrigation, water-logging drainage and conservation of water and soil and promote the development of agricultural production. In areas that are prone to salinisation-alkalisation and water-logging, measures should be taken to control and lower the groundwater level.[194]

3.6.2 Practice

This section talks about the practical situation concerning the environmental right to water in China in the two cases. This information is essential in discussing the problems relating to the environmental right to water in China.

3.6.2.1 General situation

Recognising the shortage of water resources and the deterioration of the ecological environment in some regions as being one of the principal problems in economic and social life, the Chinese government has set forth improving ecological conservation and environmental protection as one of its main targets for economic and social development, emphasising the coordination of economic development with social development.[195] In practice, the government

---

191 *Id.*, art. 22.

192 *Id.*, art. 26 (2).

193 *Id.*, art. 27 (1).

194 *Id.*, art. 25 (1).

195 Zhu, Rongji (2001) *Report on the Outline of the Tenth Five-Year Plan for the National Economic and Social Development (2001-2005)*, delivered at the Fourth Session of the Ninth National People's Congress on 5 March 2001, *Gazette of the Standing Committee of the NPC* (2001) 3, 168-179. Its unofficial English version available at

has paid close attention to solving issues of the ecological environment, taking further steps to implement the strategy of sustainable development, and stimulating coordinated economic, social and environmental development. According to the Freshwater Country Profile 2004 – China, the Chinese government has put water conservation high on its work agenda, strengthened the prevention and control of water pollution, emphasised the need to protect and make proper use of valuable resources including water in accordance with the law, improved ecological conservation and strengthened environmental protection.[196] The measures taken include revising ideas and policies, making plans for the sustainable utilisation of water resources, restructuring industries and productive forces, accelerating legislation to regulate water resource optimisation, increasing spending on water conservation, stepping up research and development to support the optimisation of water, strengthening water resource management and distribution, and strengthening groundwater monitoring and estimation.[197] Among activities organised by the Chinese government, the events of diversion waters in or for the Huanghe River, Heihe River, Tarim River, Taihu Lake and Nansihu Lake, and the construction of the South-to-North Water Diversion Project are of great value in realising the environmental rights to water.[198] Hereby, the ecological import of water for the Nansihu Lake and South-to-North Water Diversion Project are taken as examples, and will now be examined.

3.6.2.2 Case one: Ecological import of water for the Nansihu Lake[199]

As the tenth biggest freshwater lake in China and the biggest freshwater lake in northern China, with a surface area of 1152.94 km^3, the Nansihu Lake, which consists of four small lakes, *i.e.*, the Nanyang Lake, Dusan Lake, Zhaoyang

---

http://search.chinadaily.com.cn/isearch/i_textinfo.exe?dbname=cndy_printedition&listid=18027&selectword=ZHU%20RONGJI%20REPORT (last visited 31/05/2005).

196 See the National Information of UN Commission on Sustainable Development, *Freshwater Country Profile 2004 – China*, available at UN Commission on Sustainable Development webpage http://www.un.org/esa/agenda21/natlinfo/countr/china/Waterf.pdf (last visited 31/05/2005).

197 *Id.*

198 *Id.*

199 This case is summarised from the following two articles in the MWR website, (a) *Official beginning of the ecological import of water for the Nansihu Lake*, available at http://www.mwr.gov.cn/nshss/n5.htm (last visited 31/05/2005); and (b) *Background of the ecological import of water for the Nansihu Lake*, available at http://www.mwr.gov.cn/nshss/n10.htm (last visited 31/05/2005).

Lake and Weisan Lake, is rich in ecological resources and biodiversity. There are 196 types of birds (including over 30 types of National I and II Protected Types of Birds), 78 types of fish, 78 types of hydro-ecological life, 116 types of floating algae and 249 types of floating animals on this lake. All of these have formed an excellent ecological links, and the Nansihu Lake has, as a result, been defined as one of the National Natural Conservations.

Unfortunately the continuous drought in the basin due to the lower precipitation in recently years, especially in 2002 in which the annual rainfall was only 52% of the previous average year, resulted in the Lake receiving only 9.99% of the water compared with the previous year, and at the beginning of December 2002, the Lake was almost dry. Once the ecological environment has been destroyed, it is difficult to recover. In this circumstance, on 29th November 2002, the Chinese government decided to import 110 million m^3 water to the Nansihu Lake from the Changjiang River. The project started on 8th December 2002, and ended on 24th January.

3.6.2.3 Case two: South-to-North Water Diversion Project

With the economic and especially agricultural development after the establishment of the PRC, the shortage of water resources in Huang-Huai-Hai Plain in northern China began to become apparent. On 30 October 1952, Mao Zedong, then leader, stated that: "There is water in the south, while less water in the north. It should be permitted for the north to get some water from the south if it is possible".[200] Due to the massive investment needed, and the possible detrimental impact on the environment, however, the South-to-North Water Diversion Project was under survey, discussion, design and planning until September 2002.

The rapid economic and social development since the late 1970s, together with continuous severe drought, made water shortage a vital issue in the Huang-Huai-Hai Plain at the end of the last century. With per capita 501 m^3 water resources, only 20% of the average in China and 6% in the world, the area suffers from the greatest shortage of water resources in China. The situation has not only restricted social and economic development, but has also threatened the hydrological environment in the Plain. At the same time, over 94% of the flow of the Changjiang River, the longest river in China and rich in water resources with 960 billion m^3 flows per year, goes into the sea. It is therefore feasible for the Changjiang River to export some water to the north.[201] The implementation

---

[200] Commentator of People's Daily (2002) A Great Project Benefits the Future Generation, *People's Daily*, 28 December 2002.

[201] See. *Zhu Rongji was in charge of a State Council Symposium on South-to-North Water Diversion Project*, available at MWR webpage at http://www.mwr.gov.cn/nsbdzt/ldzs/zs3.htm (last visited 31/05/2005).

of this Project would undoubtedly make an impact on the environment, especially the hydrological environment. In order to minimise the detrimental impact, over 50 plans were raised, suggested and compared, and an environment impact assessment report has been completed. Finally after considering water saving, pollution control and eco-environment protection as a basis, emphasising "saving water first and transferring water later, controlling pollution first and supplying water later, protecting the environment first and using water later", the Comprehensive Plan for the South-to-North Water Diversion Project was completed and sent to the State Council for approval in early August 2002. It was adopted in September that year. Under the Comprehensive plan, there are special sections on saving water, pollution control and eco-environment protection.[202]

Under the Comprehensive Plan, three water-transfer routes will be built, involving four major river basins, *i.e.*, from south to north, the Changjiang River Basin, the Huaihe River Basin, the Huanghe River Basin, and the Haihe River Basin (see Figure 3.4). Since 27 December, the Project has begun to be implemented.

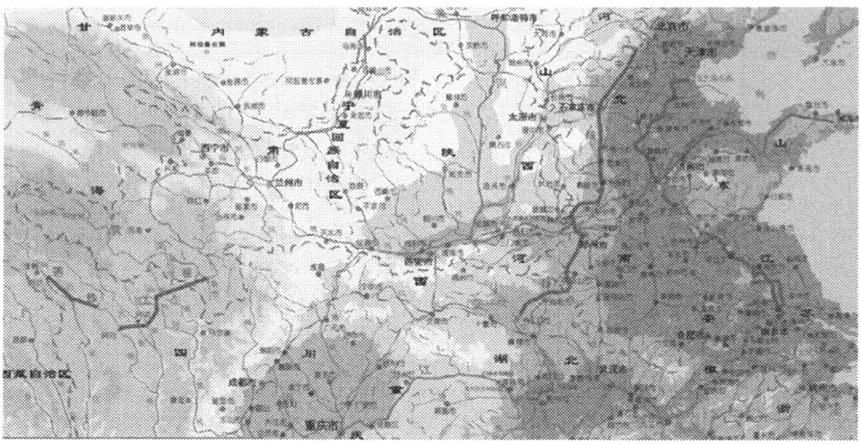

Figure 3.4. Sketch Map for the South-to-North Water Diversion Project (MWR webpage http://www.nsbd.mwr.gov.cn/nsbd/intro/brief.html).

[202] See, *The Process of Drawing Comprehensive Plan for the South-to-North Water Diversion Project*, available at MWR webpage at http://www.mwr.gov.cn/nsbdzt/jqgz/jqgz2.htm (last visited 31/05/2005).

3.6.3 Problems identified

Recognition of the environmental right to water

Although the water needs for the eco-environment are stipulated in the China Water Law 2002, as discussed above, the environmental right to water is not definitely recognised, and this makes the water needs for the environment vague to some extent. Therefore, it is necessary to make the environmental right to water clear in the China Water Law 2002.

Minimum flow for the important part of the environment

The ecological import of water for the Nansihu Lake as well as the South-to-North Water Diversion Project show that there is no effective mechanism for insuring the water needs for the eco-environment, and all measures taken depend on the decisions made by the central government case by case. Without an effective and scientific mechanism, in particular, upon which the minimum flow for the important part of the environment may be calculated, it is difficult to guarantee that the decision taken by the governments are sound.

The content of the environmental right to water

Like the human right to water, the environmental right to water is not adequately defined in/by law, or supported in practice. In particular, the environmental right to water has not obtained a proper priority among all water uses.

3.7 CONCLUSION

From the evolution of the Chinese legal system, one may conclude that the modern Chinese legal system belongs in form to the family of civil law. Under the *numerus clausus* doctrine, a fundamental theory of property law in the civil law system, the property rights, including the types and legal content of these rights, should be expressly defined by the law. Unfortunately, in reality, the term "property right" is often employed in general terms with confusing meanings, and sometimes refers to all rights that are relevant to property or of economic value. This has resulted in ownership of property or its functions not being understood properly, with goods and services often confused. This situation needs to be changed in the area of water resources management. This is because, with a lack of scientific definition, and being confused with other rights, the property right on water resources in China remains vague, and thus unenforceable. Beyond all doubt, the absence of a logical and effective system of property rights on water resources, in which the property right on water resources may be defined scientifically, will lead to a great deal of conflict and disputes. Further, this will make it difficult to undertake the integrated management of water resources.

The problems concerning the human right to water as well as the environmental right to water, such as their weak or implicit legal status, legal content and priority, also need to be resolved. Only by doing so, can the value of the social good of water resources be realised, and, further, the objective of the sustainable development and utilisation of water resources in China be reached.

In order to improve the water rights system in China, it is necessary to examine international law, since international law has lays a burden on the States, including China, in this area. Also, it is worth learning from other jurisdictions whose water law have been modernised consistently with the principle of sustainable development, and that have dealt with some aspects of water rights under a more severe water-shortage situation. In the following two chapters, the analytical approach developed in Chapter 2 will be employed to discuss the water rights both in international law and in the domestic water law of selected jurisdictions.

4

The Influence of International Law

Through analysing the relevant international legal documents, this chapter will discuss the influence of international law on domestic legislation relating to water resources and, in particular, water rights. After reviewing the evolving content of international law, including its subjects and objects, this chapter will analyse the States' obligations when enacting laws and/or policies relating to water resources, especially with respect to two aspects of water rights, *i.e.*, the human right to water and the environmental right to water. Needless to say, these obligations shall influence the legal forms and substantive content of water rights in domestic law, including Chinese water law.

The property right on water resources will not be discussed in this chapter for the following two reasons. Firstly, as a fundamental tenet of international law, each State has sovereignty over its own natural resources, including its water resources, within its jurisdiction,[1] and thus the concept of "property right" is not

[1] See, UN Resolution on Permanent Sovereignty over Natural Resources, 14 December 1962, UN Doc. A/S217 (1962) 17; UN Declaration on the Right to Development, 4

ISBN: 1843391082. Published by IWA Publishing, London, UK.

entirely relevant in international law. Secondly, the enactment of law on property right belongs to matters that lie essentially within the domestic jurisdiction of a State. For these reasons, it is difficult to find rules of international law, which directly create an obligation for States to implement in property law areas.[2]

4.1 INTRODUCTION

The term "international law", as employed here, refers to public international law. International law has a history of evolution or development in its concept as well as in its substantive legal content. Prior to the Second World War, and especially before the First World War, the subjects of international law were virtually limited to the States, *i.e.*, only the States had rights conferred and duties imposed by international law. However, with the rapid or great economic and social development after the Second World War, more and more issues have been coming under the purview of international law. Consequently, more and more international treaties not only confer rights and impose duties on the States in their relations with others State(s), but also confer rights and especially impose duties on the States to implement the content of the treaties within their jurisdictions directly or indirectly, *i.e.*, to adopt appropriate measures suiting their special conditions or circumstances to deal with relevant affairs within their jurisdictions. Further, international law confers rights and imposes duties on individuals (natural as well as legal persons), local communities or governments. Individuals, inter-government organisations, and certain territorial or political units, to various degrees, have now emerged to be considered as subjects of international law. This could be witnessed by the different versions of the book of Oppenheim's International Law. In its first three editions, the viewpoint that the State is only and exclusively the subject of international law was expressed, but gradually this book has come to recognise that objects other than the State exist so as to reflect social and international practice.[3] International law relating to human rights and sustainable development, with the

---

December 1986, UN Doc.A/RES/41/128; UN Charter of Economic Rights and Duties of States, 17 December 1984, UN Doc. A/RES/39/163.

[2] One illustration is the Article 4 (b) of UN Convention on Contracts for the International Sale of Goods, which provides that, except as otherwise expressly provided, the Convention is not concerned with "[t]he effect which the contract may have on the property in the goods sold." See art. 4 (b), UN Convention on Contracts for the International Sale of Goods, 11 April 1980, entered into forces 1 January 1988, UN Doc. A/CONF.97/18, annex I.

[3] See Jennings, R. & Watts, A., ed. (1992) *Oppenheim's International Law* (9th edition), Longman, Harlow, pp. 16-22, especially footnote 1 at page 16.

protection of the environment as core content, are areas within which international law has greatly influenced domestic law, as will be discussed in detail below.

As to the human rights aspect, the States have duties to take or provide appropriate measures, such as to adopt laws or other actions to regulate, ensure or fulfil the relevant human rights in accordance with the international law on human rights. Article 2 (2) of the International Covenant on Civil and Political Rights 1966 provides an excellent example of this. It reads as follows:

> "Where not already provided for by existing legislative or other measures, each State Party to the present Covenant undertakes to take the necessary steps, in accordance with its constitutional processes and with the provisions of the present Covenant, to adopt such laws or other measures as may be necessary to give effect to the rights recognized in the present Covenant". [4]

This can also be illustrated by Article 2 of the International Covenant on Economic, Social and Cultural Rights 1966, which provides that, "[e]ach State Party to the present Covenant undertakes to take steps, individually and through international assistance and co-operation, especially economic and technical, to the maximum of its available resources, with a view to achieving progressively the full realisation of the rights recognized in the present Covenant by all appropriate means, including particularly the adoption of legislative measures."[5]

Regarding the international law on the environment, it could be taken as an aspirational standard in domestic legislation or policy-making. In the area of sustainable development, within which the protection, development, use, conservation and administration of the environment and/or natural resources, including water resources, is a major part, international law has also influenced domestic law and/or policy to a wide extent and in great depth. As Birnie, P. W. and Boyle, A. E. point out, an environmental issue, whether it is global, regional, transboundary, domestic, or a combination of all or any of these, is addressed by international law to various extents, *i.e.*, international law addresses environmental issues at several levels, "[i]ncreasingly international law is ... addressing national or domestic environmental problems, whether through international human rights law, conservation of biological diversity, protection of natural heritage areas, or protection of sustainable development".[6]

---

4 See art. 2 (2), International Covenant on Civil and Political Rights, 16 December 1966, UN Doc. A/6316 (1966), entered into force 23 March 1976. Also available at http://www.unhchr.ch/html/menu3/b/a_ccpr.htm (last visited 31/05/2005).

5 See art. 2, International Covenant on Economic, Social and Cultural Rights, 16 December 1966, UN Doc. A/6316 (1966), entered into force 3 January 1976. Also available at http://www.unhchr.ch/html/menu3/b/treaty4_asp.htm (last visited 31/05/2005).

6 See Birnie, P. W. & Boyle, A. E. (2002) *International Law and the Environment* (2ed ed.), Oxford University Press, Oxford, pp. 6-7.

Reid, C. T. further believes that a number of principles including sustainable development "may help to give shape to the discussion and future development of the [environmental] laws", and that "although the [environmental] law continues to be structured and enforced within a national framework, the substantial rules contained in the [environmental] law are increasingly the product of international agreement at some level".[7]

The implementation of Agenda 21 is a further example of the influence of international instruments on domestic legal system. Agenda 21, together with the Rio Declaration on Environment and Development, and the Statement of Principles for the Sustainable Management of Forests, was adopted by over 178 Governments at the UN Conference on Environment and Development held in Rio de Janeiro in 1992. It outlines a wide-ranging plan of action to be adopted and implemented globally, nationally and locally by international organisations, governments at various levels, and major groups, in every area relating to the environment. Although Agenda 21 itself is a non-binding document in a strict legal concept, it has been reaffirmed and implemented by many international and/or national legal documents. For example, the "commitment to the Rio principles, the full implementation of Agenda 21 and the Programme for the Further Implementation of Agenda 21" were strongly reaffirmed by the international community at the World Summit on Sustainable Development held in Johannesburg from 26 August to 4 September 2002.[8] In accordance with Agenda 21, almost all States have adopted their relevant agenda 21 at the national level, and the relevant national or local agenda 21 has been enacted at various local government levels in many States; moreover many States have further adopted relevant legislations and/or policies to implement Agenda 21.[9]

In the case of China, its Constitution declares that "[t]he State respects and protect human rights".[10] On 27 October 1997, the government of the PRC signed the International Covenant on Economic, Social and Cultural Rights 1966, and subsequently, on 28 February 2001, the Standing Committee of the Ninth NPC approved the signature during its twentieth session.[11] As a

---

[7] See Reid, C. T., ed. (1997) *Environmental Law in Scotland* (2ed edition), W. Green, Edinburgh, pp. 2-3.

[8] See para. 1, Plan of Implementation of the World Summit on Sustainable Development, 4 September 2002, A/CONF.199/20.

[9] See *National Implementation of Agenda 21 – 2002 Country Profiles*, available at UN webpage http://www.un.org/esa/sustdev/natlinfo/cp2002.htm (last visited at 31/05/2005).

[10] See art. 33 (3), the Constitution of the PRC.

[11] See the Standing Committee of the NPC, The Decision on Ratification of International Covenant on Economic, Social and Cultural Rights, 28 February 2001, official Chinese version, *Gazette of the Standing Committee of the NPC* (2001) 2.

contracting party of the Covenant, the government of China is obliged to take appropriate measures, including legislation, to implement the human right to water, the content and origins of which will be discussed in section 4.2 of this chapter. Further, the government of China has committed itself to sustainable development, including the development, protection, conservation and utilisation of water resources, in its State practices. For example, on 25 March 1994, the State Council approved China's Agenda 21, and simultaneously a special official institution - The Administrative Center for China's Agenda 21 - was established.[12] In Chapter 14 "Conservation and Sustainable Use of Natural Resources" of China's Agenda 21, water resources is discussed. However, how to integrate the human right to water and the environmental rights into an appropriate water rights system in China, as discussed in Chapter 3, remains problematical. It is therefore necessary to analyse international law relating to the human right to water and/or the environmental right to water.

4.2 THE HUMAN RIGHT TO WATER

4.2.1 The legal status of the human right to water

As mentioned in Chapter 2, the human right to water is encompassed in human rights,[13] and it has been recognised, embodied or provided in a wide range of relevant international legal documents, including conventions, treaties, resolutions, declarations, other standards and so forth, as discussed below.

Under the African Convention on the Conservation of Nature and Natural Resources 1968, the contracting States "shall endeavour to guarantee for their populations a sufficient and continuous supply of suitable water".[14]

For human rights relating to women, the Convention on the Elimination of All Forms of Discrimination Against Women 1979 requires the States Parties take all appropriate measures to ensure that women enjoy adequate living conditions, particularly in relation to water supply.[15]

---

[12] China's Agenda 21, its Chinese version available at The Administrative Center for China's Agenda 21 webpage http://www.acca21.org.cn/cca21pa.html (last visited 31/05/2005), and its English version available at the Center's webpage http://210.72.96.19:8080/acca21/news/whitepaper.htm?classId=021101&className=China.Agenda21 (last visited 31/05/2005).

[13] See the relevant discussion in 2.4 of Chapter 2.

[14] See art. 5 (1), African Convention on the Conservation of Nature and Natural Resources 1968.

[15] See art. 14, para. 2 (h), Convention on the Elimination of All Forms of Discrimination against Women, 18 December 1979, UN Doc. A/34/46, entered into force 3 September 1981; also available at http://www.unhchr.ch/html/menu3/b/e1cedaw.htm (last visited 31/05/2005).

In the Convention on the Rights of the Child, concerning the right to the enjoyment of the highest attainable standard of health and to facilities for the treatment of illness and rehabilitation of health, the States Parties are required to take appropriate measures to pursue full implementation, and clean drinking water is among these particular measures. [16] And this is also reaffirmed in the African Charter on the Rights and Welfare of the Child 1990.[17]

In accordance with the Geneva Convention relative to the Treatment of Prisoners of War, of 1949, immediately after being captured, prisoners of war shall be supplied with sufficient potable water by the Detaining Power.[18] Thereafter, in the treatment of prisoners, the Detaining Power shall also supply prisoners of war with sufficient drinking water for living, and sufficient water for cleanness.[19] When transferring prisoners of war, the Detaining Power shall supply them with sufficient food and drinking water to keep them in good health.[20]

Under the Geneva Convention relative to the Treatment of Civilian Persons in Time of War, of 1949, in the places of internment, the Detaining Power shall provide sufficient water for internees' daily personal toilet and for washing their personal laundry.[21] Concerning food, sufficient drinking water shall be supplied to internees.[22] When the internees are transferred, the Detaining Power shall supply them with drinking water in quantity, quality and variety to maintain them in good health.[23]

In the Protocol Additional to the Geneva Conventions of 12 August 1949, and relating to the Protection of Victims of International Armed Conflicts (Protocol I), subject to some exceptions, drinking water installations and supplies in an adversary are considered as objects indispensable to the survival

---

[16] See art. 24, para. 2 (c), Convention on the Rights of the Child, 20 November 1989, UN Doc. A/44/49 (1989), entered into force 2 September 1990. Also available at http://www.unhchr.ch/html/menu3/b/k2crc.htm (last visited 31/05/2005).

[17] See art. 14 (2) (c), African Charter on the Rights and Welfare of the Child, 1990, OAU Doc. CAB/LEG/24.9/49 (1990), entered into force 29 November 1999.

[18] See art. 20, Geneva Convention relative to the Treatment of Prisoners of War, 12 August 1949, 75 UNTS 135, entered into force 21st October 1950. Also available at http://www.unhchr.ch/html/menu3/b/91.htm (last visited 31/05/2005).

[19] *Id.*, arts. 26 and 29.

[20] *Id.*, art. 46.

[21] See art. 85, Geneva Convention relative to the Protection of Civilian Persons in Time of War, 12 August 1949, 75 UNTS 287, entered into force 21st October 1950. Also available at http://www.unhchr.ch/html/menu3/b/92.htm (last visited 31/05/2005).

[22] *Id.*, art. 89.

[23] *Id.*, art. 127.

of the civilian population, and are prohibited from being attacked, destroyed, removed or rendered useless.[24]

According to the Protocol Additional to the Geneva Conventions of 12 August 1949, and Relating to the Protection of Victims of Non-International Armed Conflicts (Protocol II), to persons whose liberty has been restricted for reasons related to the armed conflict, drinking water shall be provided to the same extent as the local civilian population, and drinking water installations and supplies in an adverse Party are considered as objects indispensable to the survival of the civilian population, and are prohibited from being attacked, destroyed, removed or rendered useless.[25]

In the Mar Del Plata Action Plan of the UN Water Conference 1977, water as a "right" is recognised, and the Plan declares that, "all peoples, whatever their stage of development and their social and economic conditions, have the right to have access to drinking water in quantities and of a quality equal to their basic needs".[26] Although this plan is a non-binding document from the viewpoint of legal form, it has been implemented, and has been encompassed by other international documents. For example, the primary outcome of the conference was the launch of the International Drinking Water Supply and Sanitation Decade (1980-1990) with the slogan "Water and Sanitation for All". Later, the human right to water declared in this plan was accepted by Agenda 21,[27] and has also been recognised by the relevant international treaties[28].

The Dublin Statement on Water and Sustainable Development 1992 declares that water "is essential to sustain life".[29] Needless to say, the word "life"

---

24 See art. 54, Protocol Additional to the Geneva Conventions of 12 August 1949, and relating to the Protection of Victims of International Armed Conflicts (Protocol I), 8 June 1977, 1125 UNTS 3, entry into force 7 December 1979. Also available at http://www.unhchr.ch/html/menu3/b/93.htm (last visited 31/05/2005).

25 See arts. 5 and 14, Protocol Additional to the Geneva Conventions of 12 August 1949, and Relating to the Protection of Victims of Non-International Armed Conflicts (Protocol II), 8 June 1977, 1125 UNTS 609, entry into force 7 December 1978. Also available at http://www.unhchr.ch/html/menu3/b/94.htm (last visited 31/05/2005).

26 See Preamble, Report of the UN Water Conference, Mar del Plata, 14-25 March 1977, UN Doc. E/CONF.70/29.

27 See para. 18.47, Agenda 21.

28 For example, in the Agreement on the Action Plan for the Environmentally Sound Management of the Common Zambezi River System, the Mar Del Plata Action Plan was recognized; see para. 2 of Preamble of Agreement on the Action Plan for the Environmentally Sound Management of the Common Zambezi River System, 28 May 1987, entered into force on the same day, available at FAO webpage http://faolex.fao.org/docs/texts/bot15888.doc (last visited 31/05/2005).

29 See Principle 1, Dublin Statement on Water and Sustainable Development 1992.

employed here includes the life of human beings. In the Ministerial Declaration on 3rd World Water Forum, this notion was reconfirmed.[30]

The human right to water could also be identified in Agenda 21, in which, the substantial content of the human right to water is stated clearly in Chapter 18, which is entitled "Protection of the quality and supply of freshwater resources: application of integrated approaches to the development, management and use of water resources" and is the longest chapter of Agenda 21. For example, Paragraph 18.2 of Agenda 21 states that:

> "The general objective is *to make certain that adequate supplies of water of good quality are maintained for the entire population of this planet*, while preserving the hydrological, biological and chemical functions of ecosystems, adapting human activities within the capacity limits of nature and combating vectors of water-related diseases." [emphasis added][31]

Further, paragraph 18.8 provides that, "[i]n developing and using water resources, *priority has to be given to the satisfaction of basic needs* and the safeguarding of ecosystems." [emphasis added][32]

In the UN Convention on the Law of the Non-Navigational Uses of International Watercourses 1997, "the population dependent on the waters of the basin in each basin State" is listed as one of the factors and circumstances relevant to equitable and reasonable utilisation.[33] Further, regarding the relationship between different kinds of uses, the requirements of vital human needs shall be given special regard by watercourse States.[34]

The United Nations Economic Commission for Europe (UNECE) Protocol on Water and Health to the 1992 Convention on the Use of Transboundary Watercourses and International Lakes specifically states that "[p]arties shall, in particular, take all appropriate measures for the purpose of ensuring: (a) adequate supplies of wholesome drinking water…; (b) adequate sanitation …".[35] Further, the three central aspects of the human right to water, *i.e.*, equitable access to water, adequate quantity of water and adequate quality of water, are

---

30 See para. 1, Ministerial Declaration on 3rd World Water Forum, 23 March 2003, available at http://www.world.water-forum3.com/jp/mc/md_final.pdf (last visited 31/05/2005).

31 See para. 18.2, Agenda 21.

32 *Id.*, para. 18.8.

33 See art. 6 (1) (c), Convention on the Law of the Non-Navigational Uses of International Watercourses, 21 May 1997, UN Doc. A/51/869, not yet in force; also 36 ILM 700.

34 *Id.*, art. 10 (2).

35 See art. 4 (1), Protocol on Water and Health to the 1992 Convention on the Use of Transboundary Watercourses and International Lakes, 17 June 1999, ECOSOC doc. MP.WAT/AC.1/1999/1, entered into force 4 August 2005.

mentioned.[36] Additionally, access to water and sanitation services are reinforced in Art. 6 (1), which provides that "the Parties shall pursue the aims of: (a) access to drinking water for everyone; (b) provision of sanitation for everyone".[37]

On 17 December 1999, UN General Assembly adopted the resolution The Right to Development, which reaffirms that the right to development must be fulfilled, and "the rights to food and clean water are fundamental human rights and their promotion constitutes a moral imperative both for national Governments and for the international community".[38]

The European Council of Environmental Law (ECEL) Resolution on the Right to Water 2000, forms a definite link between human rights and water. It considers that "access to water is part of a sustainable development policy and cannot be regulated by market forces alone", and "the right to water cannot be dissociated from the right to food and the right to housing which are recognised as human right and that the right to water is also closely linked to the right to health."[39] Further, the Resolution states that "[e]ach person has the right to water in sufficient quantity and quality for his life and health".[40]

However, a particularly attention should be paid to an event happened on 26 November 2002. One of the UN millennium development goals is halving the proportion of people without sustainable access to safe drinking water.[41] In response to this, the CESCR of UN Economic and Social Council in its 29th session adopted its General Comment No. 15, which is entitled "The right to water (Articles 11 and 12 of the International Covenant on Economic, Social and Cultural Rights)". In the General Comment 15, the human right to water is raised as a substantive issue, arising out of the implementation of the Covenant, and is clearly declared a human right through the Committee's explanation. This event may be considered one of the most distinguished events in the human rights law, and the best gifts to the International Year of Freshwater, 2003.

On 23 December 2003, the UN General Assembly adopted the resolution International Decade for Action, "Water for Life", 2005-2015, which proclaims "the period from 2005 to 2015 the International Decade for Action, 'Water for Life'".[42] The goals of the Decade undoubtedly include the promotion of the

---

36 *Id.*, art. 5 (1).

37 *Id.*, art. 6 (1).

38 See para. 12, UN Resolution on the Right to Development, 17 December 1999, UN Doc. A/Res/54/175.

39 See Preamble, ECEL Resolution on the Right to Water, 28 April 2000, published in Environmental Policy and Law, 30/5 (2002).

40 *Id.*, art. 1.

41 See para. 19, United Nations Millennium Declaration, 8 September 2000, UN A/55/L.2.

42 See para. 1, UN Resolution on International Decade for Action, "Water for Life", 2005-2015, 23 December 2003, UN Doc. A/Res/58/217.

realisation of the human right to water. Firstly, the resolution emphasizes that water "is indispensable for human health and well-being".[43] Secondly, water-related goals in the international documents that are mentioned in the resolution[44] encompass the human right to water, as discussed above.

In respected literatures, the human right to water is well discussed. Early in 1999, Gleick, P. H. believed that access to a basic water requirement is a fundamental human right.[45] In the message on World Water Day 2001, UN Secretary-General Kofi Annan said, "[a]ccess to safe water is a fundamental human need and, therefore, a basic human right."[46] In The UN World Water Development Report: Water for People, Water for Life, 2003, the basic human need for water is well discussed.[47] In its publication The Right To Water, 2003, the WHO discusses the human right in great detail.[48]

It is obvious that the human right to water has a clear legal status in international law. As leading experts point out, "it can be concluded that a number of legal instruments, as well as actions, at … international levels are emerging and reinforcing the recognition of the human right to water".[49] In the following three parts, the content of the human right to water, the States' obligations under the human right to water, and the implementation of the human right to water will be discussed basically according to the General Comment 15. Readers may ask: Why the General Comment 15?

The general comments is the "interpretation of the content of human rights provisions" by the Committee on Economic, Social and Cultural Rights (CESCR).[50] The form of general comments initially originated from Article 40

---

[43] *Id.*, para. 1 of the Preamble

[44] *Id.*, para. 2.

[45] See Gleick, P.H. (1999) The Human Right to Water, *Water Policy* (1999) 1, 487-503.

[46] See UN Press Release, ACCESS TO SAFE WATER FUNDAMENTAL HUMAN NEED, BASIC HUMAN RIGHT, SAYS SECRETARY-GENERAL IN MESSAGE ON WORLD WATER DAY, UN SG/SM/7738 OBV/200, available at http://www.un.org/News/Press/docs/2001/sgsm7738.doc.htm (last visited 31/05/2005).

[47] See UNESCO, et al. (2003) *The United Nations World Water Development Report: Water for People, Water for Life*, UNESCO Publishing/Berghahn Books, Paris, pp. 120-122.

[48] See WHO (2003) *The Right To Water*.

[49] See Salman, S. M. A. & McInerney-Lankford, S. (2004) *The Human Right to Water: Legal and Policy Dimensions*, World Bank Publication, Washington, D.C., p. 83.

[50] See Office of the UN High Commissioner for Human Right webpage, *Committee on Economic, Social and Cultural Rights – General Comments*, available at http://www.ohchr.org/english/bodies/cescr/comments.htm (last visited 31/03/2004).

of International Covenant on Civil and Political Rights 1966.[51] However, "over time the general comment has become a distinct juridical instrument",[52] because of the following reasons: (a) The members of CESCR are experts with recognised competence in the field of human rights, with due consideration being given to equitable geographical distribution and to the representation of different forms of social and legal systems, and they serve in their personal capacity;[53] (b) According to the rules for drafting, the general comments shall be based on jurisprudence and other relevant documents of the other UN treaty bodies;[54] (c) "The quality of the [general] comments began to improve significantly in the late 1980's, in large measure because with the end of the Cold War it became increasingly easier within the Committee to reach the desired consensus on the manner in which different provisions of the Covenant should be interpreted";[55] and (d) "The more recent general comments tend as a rule to provide a thorough analysis of those provisions of the Covenant they seek to elucidate".[56] Therefore, Buergenthal, T., who served several terms as a member of the Committee and now serves as a judge in the International Court of Justice, concludes that, "[g]eneral comments consequently have gradually become important instrument in the lawmaking process of the Committee, independent of the reporting system".[57] In brief, although the General Comments 15 is not binding in legal form, but it extrapolate "the nature of their (the States') existing obligations".[58]

---

[51] After providing the obligation of the States Parties to submit periodic reports to the Committee, paragraph 4 of this article provides the Committee "shall study the reports" submitted by the States Parties, and "shall transit its reports, and such general comments as it may consider appropriate" to the States Parties. Further, paragraph provides that, the States Parties "may submit to the Committee observations on any comments that may be made" according to the previous paragraph. See art. 40, International Covenant on Civil and Political Rights 1966.

[52] See Alston, P. (2002) *The United Nations and Human Rights: A Critical Appraisal* (2nd ed.), Clarendon Press, Oxford, p. 711.

[53] See para. (b), Review of the composition, organisation and administrative arrangements of the Sessional Working Group of Governmental Experts on the Implementation of the International Covenant on Economic, Social and Cultural Rights, Economic and Social Council resolution 1985/17, adopted on 28 May 1985; available at http://ap.ohchr.org/documents/E/ECOSOC/resolutions/E-RES-1985-17.doc (last visited 31/05/2005).

[54] See para. I, Outline for drafting General Comments on specific rights of the ICESCR, adopted by the Committee on ESCR at its 21st session, 15 November-3 December 1999; Official Records of the Economic and Social Council, 2000, Supplement No. 2 (E/2000/22-E/C.12/1999/11 and Corr.1), annex IX; also available at http://www.unhchr.ch/html/menu2/6/cescrnote.htm (last visited 31/05/2005).

[55] See Alston, P. (2002), *supra* note 52.

[56] *Id.*

[57] *Id.*

[58] See Salman, S. M. A. & McInerney-Lankford, S. (2004), *supra* note 49, p. 5.

This work intends to review the content of the human right to water in international law, but does not aim to conduct research as to why the content of the human right to water should be other than that declared in the General Comment 15, since the author believes the Comment has been well drafted and solidly based upon the rules of international law. Consequently, it is followed.

4.2.2 The legal content of the human right to water

In general, "[t]he human right to water entitles everyone to sufficient, safe, acceptable, physically accessible and affordable water for personal and domestic uses".[59] However this description is too abstract to implement, and should be expressed in more detail. The General Comment 15 provides that:

> "The right to water contains both freedoms and entitlements. The freedoms include the right to maintain access to existing water supplies necessary for the right to water, and the right to be free from interference, such as the right to be free from arbitrary disconnections or contamination of water supplies. By contrast, the entitlements include the right to a system of water supply and management that provides equality of opportunity for people to enjoy the right to water." [60]

Further in accordance with the General Comment 15, "[t]he elements of the right to water must be adequate for human dignity, life and health",[61] and three factors, *i.e.*, availability, quality and accessibility, should apply in all circumstances, while the adequacy of water required for the right to water may vary according to different conditions.[62] In detail, the General Comment 15 sets forth the following requirements for the human right to water:

> "(a) Availability. The water supply for each person must be sufficient and continuous for personal and domestic uses. These uses ordinarily include drinking, personal sanitation, washing of clothes, food preparation, personal and household hygiene.[63] The quantity of water available for each person should correspond to World Health Organisation (WHO) guidelines. Some individuals

---

[59] See para. 2, General Comment 15.

[60] *Id.*, para. 10.

[61] *Id.*, para. 11.

[62] *Id.*, para. 12.

[63] Note 13 of the General Comment 15 gives explanations to the terms of drinking, personal sanitation, food preparation, and personal and household hygiene that, "[I]n this context, 'drinking' means water for consumption through beverages and foodstuffs. 'Personal sanitation' means disposal of human excreta. Water is necessary for personal sanitation where water-based means are adopted. 'Food preparation' includes food hygiene and preparation of food stuffs, whether water is incorporated into, or comes into contact with, food. 'Personal and household hygiene' means personal cleanliness and hygiene of the household environment".

and groups may also require additional water due to health, climate, and work conditions;

(b) Quality. The water required for each personal or domestic use must be safe, therefore free from micro-organisms, chemical substances and radiological hazards that constitute a threat to a person's health. Furthermore, water should be of an acceptable colour, odour and taste for each personal or domestic use.

(c) Accessibility. Water and water facilities and services have to be accessible to everyone without discrimination, within the jurisdiction of the State party. Accessibility has four overlapping dimensions:

(i) Physical accessibility: water, and adequate water facilities and services, must be within safe physical reach for all sections of the population. Sufficient, safe and acceptable water must be accessible within, or in the immediate vicinity, of each household, educational institution and workplace. All water facilities and services must be of sufficient quality, culturally appropriate and sensitive to gender, life-cycle and privacy requirements. Physical security should not be threatened during access to water facilities and services;

(ii) Economic accessibility: Water, and water facilities and services, must be affordable for all. The direct and indirect costs and charges associated with securing water must be affordable, and must not compromise or threaten the realisation of other Covenant rights;

(iii) Non-discrimination: Water and water facilities and services must be accessible to all, including the most vulnerable or marginalized sections of the population, in law and in fact, without discrimination on any of the prohibited grounds; and

(iv) Information accessibility: accessibility includes the right to seek, receive and impart information concerning water issues." [64]

In the General Comment 15, special emphasis is given to non-discrimination and equality in the broad application of the human right to water.[65]

4.2.3 The obligations of the States

Under the International Covenant on Economic, Social and Cultural Rights 1966, the States are required to undertake certain obligations to realise and acknowledge human rights, and the human right to water should not be excluded. In accordance with the General Comment 15, the States' obligations are discussed at four levels, *i.e.*, general legal obligations, specific legal obligations, international obligations relating to the shared international water resources and core obligations. However, international obligations lie beyond the scope of this work. Instead, the specific legal content of the human right to

---

[64] See para. 12, General Comment 15.

[65] *Id.*, paras. 13-16.

water will be examined in some detail, with a view to determining whether there are lessons to be learned.

4.2.3.1 General legal obligations

The States have, (a) relevant immediate obligations, *e.g.*, the guarantee that the human right to water will be implemented without any discrimination, and (b) the obligations to take deliberate, concrete steps targeted towards the full fulfilment of the human right to water.[66]

4.2.3.2 Specific legal obligations

Specific legal obligations consist of three forms: (a) The obligation to respect, under which the States must "refrain from interfering directly or indirectly with the enjoyment of the right to water".[67] This includes "refraining from engaging in any practice or activity that denies or limits equal access to adequate water; arbitrarily interfering with customary or traditional arrangements for water allocation; unlawfully diminishing or polluting water, for example through waste from State-owned facilities or through use and testing of weapons; and limiting access to, or destroying, water services and infrastructure as a punitive measure, for example, during armed conflicts in violation of international humanitarian law".[68] (b) The obligation to protect, under which the States must "prevent third parties from interfering in any way with the enjoyment of the right to water".[69] This includes, "adopting the necessary and effective legislative and other measures to restrain, for example, third parties from denying equal access to adequate water; and polluting and inequitably extracting from water resources, including natural sources, wells and other water distribution systems".[70] (c) The obligation to fulfil, *i.e.*, "to take positive measures to assist individuals and communities to enjoy the right" (the obligation to facilitate), "to take steps to ensure that there is appropriate education concerning the hygienic use of water, protection of water sources and methods to minimize water wastage" (the obligation to promote), "to fulfil (provide) the right when individuals or a group are unable, for reasons beyond their control, to realize that right themselves by the means at their disposal".[71]

In short, the specific legal obligation of the States relating to the human right to water includes the obligations to respect, to protect, and to fulfil.

---

[66] *Id.*, para. 17.
[67] *Id.*
[68] *Id.*
[69] *Id.*, para. 23.
[70] *Id.*
[71] *Id.*, para. 25.

4.2.3.3 Core obligations

The States are under the following core obligations "to ensure the satisfaction of, at the very least, minimum essential levels" of the human right to water:

> "(a) To ensure access to the minimum essential amount of water, that is sufficient and safe for personal and domestic uses to prevent disease;
>
> (b) To ensure the right of access to water and water facilities and services on a non-discriminatory basis, especially for disadvantaged or marginalized groups;
>
> (c) To ensure physical access to water facilities or services that provide sufficient, safe and regular water; that have a sufficient number of water outlets to avoid prohibitive waiting times; and that are at a reasonable distance from the household;
>
> (d) To ensure personal security is not threatened when having to physically access to water;
>
> (e) To ensure equitable distribution of all available water facilities and services;
>
> (f) To adopt and implement a national water strategy and plan of action addressing the whole population; the strategy and plan of action should be devised, and periodically reviewed, on the basis of a participatory and transparent process; it should include methods, such as right to water indicators and benchmarks, by which progress can be closely monitored; the process by which the strategy and plan of action are devised, as well as their content, shall give particular attention to all disadvantaged or marginalized groups;
>
> (g) To monitor the extent of the realization, or the non-realization, of the right to water;
>
> (h) To adopt relatively low-cost targeted water programmes to protect vulnerable and marginalized groups;
>
> (i) To take measures to prevent, treat and control diseases linked to water, in particular ensuring access to adequate sanitation." [72]

In summary, the General Comment 15 sets forth in some detail the core requirements for the States to deal with the human right to water. These should be useful for all States Parties to the International Covenant on Economic, Social and Cultural Rights 1966, including China.

4.2.4 Implementation at national level

The International Covenant on Economic, Social and Cultural Rights 1966 requires the States to employ "all appropriate means, including particularly the adoption of legislative measures" in the implementation of their Covenant

[72] *Id.*, para. 37.

obligations.[73] It is submitted that the States' obligations concerning the human right to water should follow the same approach. In the implementation of the human right to water at national law, the following aspects should be taken into consideration.

4.2.4.1 Formulation of legislation, strategies and policies

Where there is existing legislation, strategies and/or policies, the States should evaluate them in order to ensure that they are compatible with their obligations under the human right to water, and repeal, amend or change them if they are contradictory with the requests under the International Covenant on Economic, Social and Cultural Rights 1966.[74] When there is no legislation, strategies and policies, the States should undertake necessary measures to formulate ones to introduce such principles, standards or procedures.[75]

4.2.4.2 Indicators and benchmarks

In order to implement the human right to water effectively and evaluate this implementation properly, indicators and benchmarks should be established by every State. By being recognised in their national water strategies or plans of action, "[i]ndicators should address the different components of adequate water (such as sufficiency, safety and acceptability, affordability and physical accessibility), be disaggregated by the prohibited grounds of discrimination, and cover all persons residing in the State's territorial jurisdiction or under their control".[76] It is vital that the implementation of the human right to water is monitored to be systematically, and indicators play a significant role in this task.

4.2.4.3 Remedies and accountability

In implementing the human right to water, effective remedies and an accountability mechanism are essential. A relevant mechanism should therefore be established, where there is not one, or improved, where there is an existing one, in order to ensure the realisation of the human right to water. Firstly, "[a]ny persons or groups who have been denied their right to water should have access to effective judicial or other appropriate remedies at both national and international levels".[77] This access to justice must be assured.

---

73 See art. 2 (1), International Covenant on Economic, Social and Cultural Rights 1966.
74 See para. 46, General Comment 15.
75 *Id.*, paras. 47-52.
76 *Id.*, para. 53.
77 *Id.*, para. 55.

Secondly, "[b]efore any action that interferes with an individual's human right to water is carried out by the State, or by any other third party, the relevant authorities must ensure that such actions are performed in a manner warranted by law, compatible with the Covenant".[78] This is a wise suggestion. If actions interfering with the human rights to water are permitted to be carried out in the absence of a law that is consistent with the Covenant, the protection of the human right to water cannot be guaranteed successfully.

Thirdly, the incorporation in the domestic legislation of international instruments recognising the human right to water should be encouraged in all cases, because this incorporation can considerably improve the scope and effectiveness of the remedial instruments, and "enables courts to adjudicate violations of the right to water, or at least the core obligations, by direct reference to the Covenant".[79]

Fourthly, the States should support judges, adjudicators and members of the legal occupation in giving greater attention to violations of the human right to water in exercising their functions.[80]

Finally, the States "should respect, protect, facilitate and promote the work of human rights advocates and other members of civil society with a view to assisting vulnerable or marginalized groups in the realization of their right to water".[81]

From the above discussions, one can see the human right to water has been recognised in international law. Further, the realisation of the human right to water requires the States to take appropriate measures. The author believes that, domestic law should play a key role in this matter. That is to say, the legal status, content, and the priority among all water uses of the human right to water, should be clearly defined in law.

4.3 THE ENVIRONMENTAL RIGHT TO WATER

In a research on the environmental flows carried out by experts from the IUCN - The World Conservation Union (hereinafter "IUCN"), two steps are suggested being taken for creating a national policy and legal framework on the environmental flows.[82] "The first step is to determine what" international "instruments exist that may have an influence on the policy decisions and actions taken at the national level".[83] The second step involves examining the

---

[78] *Id.*, para. 56.
[79] *Id.*, para. 57.
[80] *Id.*, para. 58.
[81] *Id.*, para. 59.
[82] See Dyson, M., et al. ed. (2003) *Flow: The essentials of environmental flows*, IUCN, Cambridge, pp. 73, 82-83.
[83] *Id.*

situation in a particular country, especially checking the international documents relating to that country.[84] This approach can be employed in this work. As a first step, the content of international legally binding and soft law instruments relating to the environmental right to water will now be identified. For the second step, the international documents relating to the environmental right to water regarding China will be examined in Chapter 6.

4.3.1 The legal status of the environmental right to water

As mentioned in Chapter 2, due to the vital role of water in relation to the environment, it is logical to conclude that there should be an environmental right to water, and, in fact, this right has been embodied in legal documents of many jurisdictions.[85] Regarding the legal status of the environmental right to water in international law, since the 1970's, particularly since the UN Conference on the Human Environment (Stockholm, 5-16 June 1972), many comprehensive as well as special international legal documents dealing with water resources have been adopted, and the environmental right to water has been stipulated. These will be analysed below.

In the Convention on Wetlands of International Importance especially as Waterfowl Habitat 1971 (hereinafter "Ramsar Convention"), the interdependence between the human beings and the environment is recognised, and the fundamental ecological functions of wetlands are considered.[86] Further, the Convention sets up relevant duties for the Contracting Parties to fulfil.[87] Needless to say, without water, wetlands cannot survive. It is therefore logical to conclude that the international important wetlands, the key element of the environment, have a right to access water.

In order to protect the cultural and natural heritage of outstanding universal value, the Convention concerning the Protection of the World Cultural and Natural Heritage 1972 (hereinafter "World Heritage Convention") was adopted.[88] Regarding the scope of the natural heritage, the Convention provides that:

---

[84] *Id.*

[85] See the relevant discussion in 2.5 of Chapter 2.

[86] See paras. 2-3 of Preamble, Convention on Wetlands of International Importance especially as Waterfowl Habitat, 2 February 1971, 996 UNTS 245, entered into force 21 December 1975, as amended by the Protocol of 3 December 1982 and the Amendments of 28 May 1987; an official full version available at http://www.ramsar.org/key_conv_e.htm (last visited 31/05/2005).

[87] *Id.*, arts. 2-8.

[88] See Preamble, Convention concerning the Protection of the World Cultural and Natural Heritage, 16 November 1972, 11 ILM 1358, entered into force 17 December 1975.

> "For the purposes of this Convention, the following shall be considered as 'natural heritage':
>
> natural features consisting of physical and biological formations or groups of such formations, which are of outstanding universal value from the aesthetic or scientific point of view;
> geological and physiographical formations and precisely delineated areas which constitute the habitat of threatened species of animals and plants of outstanding universal value from the point of view of science or conservation;
>
> natural sites or precisely delineated natural areas of outstanding universal value from the point of view of science, conservation or natural beauty." [89]

It is natural that, the important parts of the environment, including lakes, rivers, or water dependent ecosystems, can be listed as the natural heritage, and this is also the fact.[90] Although this World Heritage Convention and the Ramsar Convention operate on the basis of voluntary list, once a wetland has been listed as a wetland of international importance, or a lake, river or water dependent ecosystem has been listed as a natural heritage,[91] it then is under both national and international protection.

In the Declaration of the UN Conference on the Human Environment 1972, Principle 2 declares that:

> "The natural resources of the earth, including the air, water, land, flora and fauna and especially representative samples of natural ecosystems, must be safeguarded for the benefit of present and future generations through careful planning or management, as appropriate." [92]

Flora and fauna live in and/or on water, and would be difficult to preserve without water. It is clear that the environmental right to water is recognised, although implicitly.

As to the Convention on the Conservation of Migratory Species of Wild Animals, [93] experts from IUCN consider it "indirectly relevant for environmental flows management", because it "can be useful for the conservation of environmental flows when rivers and wetlands constitute the

---

89 *Id.*, art. 2.

90 In accordance with the Ramsar Convention, the List is determined by the State itself; see art. 2, Ramsar Convention. According to the World Heritage Convention, each State Party may identify and delineate the sites, and propose a sites in the World Heritage List; following a decision of the World Heritage Committee, a proposed site may be listed; see arts. 3 and 11, World Heritage Convention.

91 Such as Jiuzhaigou Valley Scenic and Historic Interest Area, Wulingyuan Scenic and Historic Interest Area and Mount Huangshan. See UNESCO, *World Heritage List*, available at UNESCO webpage http://whc.unesco.org/pg.cfm?cid=31(last visited 31/05/2005).

92 See Principle 2, Declaration of the UN Conference on the Human Environment 1972.

93 The Convention on the Conservation of Migratory Species of Wild Animals, 23 June 1979, 19 ILM 15, entered into force 1 November 1983.

habitat of protected species and the maintenance of water flows is necessary to ensure the survival of a migratory species".[94]

Among the four Dublin principles of The Dublin Statement on Water and Sustainable Development 1992, the first is directly related to the substantial content of the environmental right to water. It declares that fresh water is "essential to ... the environment".[95] Water maintains all life on the Earth; therefore, effective management of water resources demands a holistic approach, *i.e.*, considering social and economic development as well as the protection of natural ecosystems. In other words, the environment should be protected for its own or intrinsic value.

According to the Convention on the Protection and Use of Transboundary Watercourses and International Lakes 1992, the contracting parties are under a duty to take all appropriate measures "[t]o ensure that transboundary waters are used with the aim of ecologically sound and rational water management, conservation of water resources and environmental protection", "[t]o ensure conservation and, where necessary, restoration of ecosystems".[96]

In the UN Framework Convention on Climate Change 1992, the environmental right to water is also provided. The Convention states that its objective is to achieve "stabilization of greenhouse gas concentrations in the atmosphere at a level that would prevent dangerous anthropogenic interference with the climate system. Such a level should be achieved within a time frame sufficient to allow ecosystems to adapt naturally to climate change...".[97]

From the relevant principles of the Rio Declaration on Environment and Development, *i.e.*, Principles 1, 4, 7, 23 and 25, one also can conclude that the environmental right to water is provided implicitly, since water is an indispensable component of the environment.[98] This can be identified in

---

94 See Dyson, M., et al. ed., *supra* note 82, p. 77.

95 See principle 1, The Dublin Statement on Water and Sustainable Development 1992. It reads as follows:

> "Fresh water is a finite and vulnerable resource, essential to sustain life, development and the environment."

96 See art. 2 (2) (b) and (d), Convention on the Protection and Use of Transboundary Watercourses and International Lakes, 17 March 1992, 1936 UNTS 269, entered into force 6 October 1996.

97 See art. 2, UN Framework Convention on Climate Change, 9 May 1992, 1771 UNTS 107, entered into force 21 March 1994; 31 ILM 849 (1992).

98 See Principles 1, 3, 4, 7, 23 and 25, Rio Declaration on Environment and Development. In accordance with Principle 1, although human beings "are entitled to a healthy and productive life", such a life should be "in harmony with nature". And Principle 4 reads as follows:

Agenda 21, in which the substantial content of the environmental right to water is stated clearly in Chapter 18. For example, Paragraph 18.2 of Agenda 21 states that:

> "The general objective is to make certain that adequate supplies of water of good quality are maintained for the entire population of this planet, *while preserving the hydrological, biological and chemical functions of ecosystems*, adapting human activities *within the capacity limits of nature* and combating vectors of water-related diseases." [emphasis added][99]

Considering the functioning of aquatic ecosystems and the perenniality of the resource, to satisfy and reconcile needs for water in human activities, paragraph 18.8 provides that:

> "In developing and using water resources, priority has to be given to the satisfaction of basic needs and *the safeguarding of ecosystems*." [emphasis added][100]

The title of Part C of Chapter 18 is "Protection of water resources, water quality and aquatic ecosystems". Under this part, paragraph 18.36 stipulates that freshwater management should be holistic and "based on a balanced consideration of the needs of people and the environment."[101] Paragraph 18.38(a) further stipulates that:

> "Maintenance of ecosystem integrity, according to a management principle of preserving aquatic ecosystems, including living resources, and of effectively protecting them from any form of degradation on a drainage basin basis." [102]

And paragraph 18.39(g) requires the States "[t]o adopt an integrated approach to environmentally sustainable management of water resources, including the protection of aquatic ecosystems and freshwater living resources".[103]

> "In order to achieve sustainable development, environmental protection shall constitute an integral part of the development process and cannot be considered in isolation from it."

Principle 7 declares that, "States shall cooperate in a spirit of global partnership to conserve, protect and restore the health and integrity of the Earth's ecosystem". Further, Principle 23 emphasises that, "[t]he environment and natural resources of people under oppression, domination and occupation shall be protected".

Regarding the relations among peace, development and environmental protection, Principle 26 states that, "[p]eace, development and environmental protection are interdependent and indivisible".

99 See para. 18.2, Agenda 21.

100 *Id.*, para. 18.8.

101 *Id.*, para. 18.36.

102 *Id.*, para. 18.38(a).

103 *Id.*, para. 18.39(g).

In brief, according the Rio Declaration on Environment and Development, and Agenda 21, the environment has a right to water, and the safeguarding of ecosystems has to be given priority in development and utilisation of water resources.

Under the UN Convention on Biological Diversity 1992, the variability among living organisms from aquatic ecosystems and the ecological complexes of which they are part is one component of biological diversity,[104] and the intrinsic value of the biological diversity is acknowledged.[105] It is logical to conclude that the ecosystem has a right to get the minimum amount of water to sustain itself.

Among the objects of the Agreement on the Cooperation for the Sustainable Development of the Mekong River Basin, 1995, there are three objects directly concerning the environmental right to water, *i.e.*, the protection of the environment and ecological balance, the maintenance of flows on the mainstream, and the prevention and cessation of harmful effects that might occur to the environment.[106]

---

[104] See article 2, UN Convention on Biological Diversity, 5 June 1992, 1760 UNTS 79, entered into force 29 December 1993. Also available at http://www.biodiv.org/convention/articles.asp (last visited 31/05/2005).

[105] *Id.*, paras. 1-2 of the Preamble.

[106] See arts. 3, 6-7, Agreement on the Cooperation for the Sustainable Development of the Mekong River Basin, 5 April 1995, entered into force 5 April 1995, available at Mekong River Commission webpage http://www.mrcmekong.org/pdf/95%20Agreement.pdf (last visited 31/05/2005); also 34 ILM 864 (1995). Article 3 provides that, "[t]o protect the environment, natural resources, aquatic life and conditions, and ecological balance of the Mekong River Basin from pollution or other harmful effects resulting from any development plans and uses of water and related resources in the Basin".

Article 6 reads as follows:

> "To cooperate in the maintenance of the flows on the mainstream from diversions, storage releases, or other actions of a permanent nature; except in the cases of historically severe droughts and/or floods:
>
> A. Of not less than the acceptable minimum monthly natural flow during each month of the dry season;
>
> B. To enable the acceptable natural reverse flow of the Tonle Sap to take place during the wet season; and.
>
> C. To prevent average daily peak flows greater than what naturally occur on the average during the flood season.

Article 6 (1) (a) of the UN Convention on the Law of the Non-Navigational Uses of International Watercourses 1997 provides that, in determining what is an equitable and reasonable utilisation of an international watercourse, "[g]eographic, hydrographic, hydrological, climatic, ecological and other factors of a natural character" should be taken into consideration with all other relevant factors and circumstances.[107] Further, Article 20 requires "[w]atercourse States shall, individually and, where appropriate, jointly, protect and preserve the ecosystem of international watercourses".[108] As one expert points out, there is an "emerging obligation (for the watercourse States) to protect the ecosystems of international watercourses".[109]

In the Ministerial Declaration of The Hague on Water Security in the 21st Century, agreed on Wednesday 22 March 2000, in The Hague, Netherlands, the ministers recognised that, to provide water security "means ensuring that freshwater, coastal and related ecosystems are protected and improved".[110] Further, seven challenges were adopted as the basis for future action. Among these challenges, two are directly to the content of the environmental right to water, *i.e.*, protecting ecosystems – ensuring their integrity via sustainable water management, and valuing water – to manage water in the light of its different values, including environmental value.[111]

In the UN Millennium Declaration 2000, under Part IV which is entitled "Protecting our common environment", paragraph 22 states that, "[w]e reaffirm

---

> The Joint Committee shall adopt guidelines for the locations and levels of the flows, and monitor and take action necessary for their maintenance as provided in Article 26."

And article 7 stipulates that:

> "To make every effort to avoid, minimize and mitigate harmful effects that might occur to the environment, especially the water quantity and quality, the aquatic (eco-system) conditions, and ecological balance of the river system, from the development and use of the Mekong River Basin water resources or discharge of wastes and return flows. Where one or more States is notified with proper and valid evidence that it is causing substantial damage to one or more riparians from the use of and/or discharge to water of the Mekong River, that State or States shall cease immediately the alleged cause of harm until such cause of harm is determined in accordance with Article 8."

107 See art. 6 (1) (a), Convention on the Law of the Non-Navigational Uses of International Watercourses 1997.

108 *Id.*, art. 20.

109 See McCaffrey, S. C. (2001) *The Law of International Watercourses: Non-Navigational Uses*, Oxford University Press, Oxford, p. 396.

110 See para. 1, Ministerial Declaration of the Hague on Water Security in the 21st Century, available at http://www.worldwaterforum.net/Ministerial/declaration.html (last visited 31/05/2005).

111 *Id.*, para. 3.

our support for the principles of sustainable development, including those set out in Agenda 21, agreed upon at the United Nations Conference on Environment and Development".[112] According to paragraph 23, under a new ethic of conservation and stewardship, each State would resolve "[t]o stop the unsustainable exploitation of water resources by developing water management strategies at the regional, national and local levels, which promote both equitable access and adequate supplies".[113]

Under the Revised Protocol on Shared Watercourses 2000, the State Parties in the Southern African Development Community (SADC), shall individually and/or jointly, (i) protect and preserve the ecosystems of, (ii) prevent, reduce and control the pollution and environmental degradation of, (iii) prevent the introduction of harmful species (alien or new) into, and (iv) protect and preserve the aquatic environment of a shared watercourses.[114]

In the Johannesburg Declaration on Sustainable Development 2002, the States reaffirm their commitment to sustainable development, and to "undertake to strengthen and improve governance at all levels for the effective implementation of Agenda 21, the Millennium development goals and the Plan of Implementation of the Summit".[115]

In the Ministerial Declaration adopted on 23 March 2003 at the Third Water Forum, Tokyo, Japan, on the basis of recognition of "the need to intensify water pollution prevention in order to reduce hazards to health and the environment and to protect ecosystems, including control of invasive species",[116] the ministers stated that, "[t]o ensure a sustainable water supply of good quality",[117] they "should protect and use in a sustainable manner the ecosystems that naturally capture, filter, store, and release water, such as rivers, wetlands, forests, and soils ",[118] and further urged "countries to review and, when necessary, to establish appropriate legislative frameworks for the protection and sustainable use of water resources and for water pollution prevention".[119]

In other international documents, the environmental right to water can also be identified.[120]

---

112 See para. 22, UN Millennium Declaration.

113 *Id.*, para. 23.

114 See art. 4 (2) (a), (b) (i), (c) and (d), Revised Protocol on Shared Watercourses, 7 August 2000, not yet in force, available at http://www.iucnrosa.org.zw/elisa/SADC-Protocols/rivised_shared_watercourse_protocol.html (last visited 31/05/2005).

115 See para. 30, Johannesburg Declaration on Sustainable Development.

116 See para. 23, Ministerial Declaration on 3rd World Water Forum 2003.

117 *Id.*, para. 24.

118 *Id.*

119 *Id.*, para. 24.

120 See the following discussion in 4.3.2, 4.3.3 and 4.3.4 of this chapter.

In the UN World Water Development Report: Water for People, Water for Life, it stated that all the goals most relevant to water among the Millemium Development Goals for 2015 need to be achieved "while protecting the environment from further degradation".[121] Protecting ecosystems for people and the planet is listed as the second challenge to life and well-being on the Earth, and human beings "have come to accept two important concepts in the past decade: firstly, that ecosystems not only have their own intrinsic value, but also provide humankind with essential services; secondly, that the sustainability of water resources requires participatory, ecosystem-based management".[122]

According to World Bank's Water Resources Sector Strategy (2003), one of the three fundamental principles in modern water resources management is the ecological principle, and "the environment is, of course, a special 'water-using sector' in that most environmental concerns are a central part of overall water resources management, and not part of a distinct water-using sector".[123]

As demonstrated above, healthy environment is a cornerstone of sustainable development, and "[i]nternational law has now progressed to point that it protects" this value.[124] In a word, the environmental right to water has been recognised, embodied and reflected in numerous international legal documents, although implicitly in some cases.

4.3.2 The legal content of the environmental right to water

From the content of the international documents employed in the discussion on the legal status of the environmental right to water, the author would like to conclude that the environmental right to water entitles the environment or nature to the minimum amount of water with appropriate quality to sustain the environment, nature or ecosystem in a healthy order or status.

In the following two parts, the States' obligations under the environmental right to water, and the implementation of the environmental right to water will be examined basically according to Agenda 21. Readers may ask: Why Agenda 21?

Firstly, Agenda 21 indicates a high degree of international consensus, because it was adopted by over 178 Governments in 1992. Further, the full execution of Agenda 21, the Programme for Further Implementation of Agenda 21 and the Commitments to the Rio principles, were strongly reaffirmed at the

---

121 See UNESCO, et al. (2003) *The United Nations World Water Development Report: Water for People, Water for Life (Executive Summary)*, UNESCO Publishing/Berghahn Books, Paris, p. 6.

122 *Id.*, p. 13.

123 See World Bank (2003), *Water Resources Sector Strategy: Strategic Directions for World Bank Engagement*, World Bank Publication, Washington, D.C., pp. v, 20.

124 See McCaffrey, S. C., *supra* note 109, p. 94.

World Summit on Sustainable Development held in Johannesburg in 2002. Secondly, in order to ensure effective follow-up of documents adopted by the UN Conference on Environment and Development, the UN General Assembly adopted the resolution of Institutional arrangements to follow up the United Nations Conference on Environment and Development,[125] in accordance with which, The UN Commission on Sustainable Development was created in December 1992. As a functional commission of the UN Economic and Social Council, the UN Commission on Sustainable Development monitors and reports on implementation of the Earth Summit agreements at the local, national, regional and international levels. Thirdly, as a comprehensive plan of action to be taken globally, nationally and locally by the organizations of the United Nations System, governments, and major groups in every area in which humans impact on the environment, it has been enforced or implemented at different levels. For example, Agenda 21 is recalled and kept in remembrance in the Agreement to Promote Compliance with International Conservation and Management Measures by Fishing Vessels on the High Seas 1993.[126] In Southern Africa Development Community, the Protocol on Shared Watercourse 1995 as well as in the Revised Protocol on Shared Watercourses 2000, the relevant provisions relating to the shared watercourses of Agenda 21 are recognised.[127] Under the African Convention on the Conservation of Nature and Natural Resources 2003, "the need to continue furthering the principles of the Stockholm Declaration, to contribute to the implementation of the Rio Declaration and of Agenda 21" is conscious.[128]

---

125 Institutional arrangements to follow up the United Nations Conference on Environment and Development, 22 December 1992, UN Doc A/RES/47/191; available at http://www.un.org/documents/ga/res/47/ares47-191.htm (last visited 31/05/2005).

126 See paras. 4 and 6 of Preamble, Agreement to Promote Compliance with International Conservation and Management Measures by Fishing Vessels on the High Seas, 24 November 1993, entered into force 24 April 2003, available at FAO webpage http://www.fao.org/legal/treaties/012t-e.htm (last visited 31/05/2005).

127 See para. 2 of Preamble, Protocol on Shared Watercourse, 28 August 1995, entered into force 29 September 1998, available at FAO webpage http://www.fao.org/docrep/w7414b/w7414b0n.htm (last visited 31/05/2005). See para. 2 of Preamble, Revised Protocol on Shared Watercourses 2000.

128 See para. 10 of Preamble, African Convention on the Conservation of Nature and Natural Resources, 11 July 2003, not yet in force, available at http://www.africa-union.org/Official_documents/Treaties_%20Conventions_%20Protocols/nature%20and%20natural%20recesource.pdf (last visited 31/05/2005).

4.3.3 The obligations of the States

In order to realise the environmental right to water to a reasonable degree, international law imposes certain obligations on the States. Under these obligations, the States are required to take appropriate steps to realise the environmental right to water.[129] The obligations of the States relating to the environmental right to water, like that relating to the human right to water, can be discussed at four levels, *i.e.*, general legal obligations, specific legal obligations, international obligations concerning the shared international water resources, and core obligations. These will be discussed in more detail below, excluding international obligations, because international obligations lie outside the scope of this work.

4.3.3.1 General legal obligations

The environment has a right to get the minimum amount of proper water required to sustain itself, including all of its relevant ecosystems, and the present human beings are under a duty as well as an entitlement to realise the above right. However, situations relating to water resources differ from State to State, and from region to region, and there exist different economic and social development levels. Therefore, the restrictions due to the limits of available resources of different States must be acknowledged. Nevertheless, the States shall have corresponding obligations in relation to the environmental right to water. The States have the immediate obligations, *e.g.*, to recognise that the environment has a right to a minimum of water to sustain itself in a healthy status, as well as obligations to take appropriate steps towards the reasonable realisation of the environmental right to water. The steps must be conscious, actual and targeted towards the realisation of the environmental right to water, although the States could adopt these steps "according to their capacity and available resources" in accordance with Agenda 21.[130]

According to paragraph 18.40 of Agenda 21, all States could implement relevant special activities to realise the substantial content of the environmental right to water. Among them, the following are of most importance:

(1) in water resources protection and conservation, "[p]reparation of national plans for water resources protection and conservation",

---

[129] In general, see paras. 18.12–18.22, Agenda 21. In special areas, see the relevant discussion on the international documents in 4.3.3 of this chapter. For example, Article 7(1) (a) of African Convention on the Conservation of Nature and Natural Resources 2003 provides that, the Parties shall take measures designed to "maintain water-based essential ecological processes as well as to protect human health against pollutants and water-borne diseases". See art. 7(1) (a), African Convention on the Conservation of Nature and Natural Resources 2003.

[130] See paras. 18.11, 18.12, 18.26, 18.27, 18.39, 18.40, 18.50, 18.58, 18.59, Agenda 21.

"[r]ehabilitation of important, but degraded catchment areas, particularly on small islands".[131]

(2) In water pollution prevention and control, "[m]andatory environmental impact assessment of all major water resource development projects potentially impairing water quality and aquatic ecosystems, combined with the delineation of appropriate remedial measures and a strengthened control of new industrial installations, solid waste landfills and infrastructure development projects".[132]

(3) in protection of aquatic ecosystems, "[r]ehabilitation of polluted and degraded water bodies to restore aquatic habitats and ecosystems", "[c]onservation and protection of wetlands (owing to their ecological and habitat importance for many species), taking into account social and economic factors", "[c]ontrol of noxious aquatic species that may destroy some other water species".[133]

(4) in protection of freshwater living resources, "[p]rotection of ecosystem from pollution and degradation for the development of freshwater aquaculture projects".[134]

To conclude the above discussions, the general legal obligations of the States could be summarised as, (a) relevant immediate obligations, *e.g.*, the guarantee that the environmental right to water will be implemented, and (b) the obligations to take deliberate, concrete steps targeted towards the fulfilment of the environmental right to water suitable to their capacity and available resources.

4.3.3.2 Specific legal obligations

Like the human right to water, the environmental right to water imposes three types of obligations on the States, *i.e.*, obligations to respect, obligations to protect and obligations to fulfil. These shall be examined next.

Obligations to respect The obligation to respect requires the States to refrain from interfering directly or indirectly with the enjoyment of the environmental right to water. The obligation includes, *inter alia*, refraining from engaging in any practice or activity that denies the environment's access to minimum water, unlawfully diminishing or polluting water.[135]

---

131 *Id.*, para. 18.40 (a) (iii) and (iv).

132 *Id.*, para. 18.40 (b) (v).

133 *Id.*, para. 18.40 (e) (i), (iii) and (iv). Also, arts. 2-8, Ramsar Convention.

134 See para. 18.40(f) (ii), Agenda 21.

135 *Id.*, para. 18.39 (f). See also para 4 of the Preamble, Agreement on the Cooperation for the Sustainable Development of the Mekong River Basin 1995; African Convention on the Conservation of Nature and Natural Resources 2003.

Obligations to protect The obligation to protect requires the States to protect the environmental right to water, particularly prevent third parties from interfering in any way with the enjoyment of the environmental right to water. Third parties include individuals, groups, corporations and other entities as well as agents acting under their authority. The obligation includes, *inter alia*, (a) protecting the environmental right to water,[136] and (b) adopting the necessary and effective legislative and other measures to restrain, for example, third parties from denying the environment's access to minimum water, and polluting water resources and over extracting water from water sources.[137]

Obligations to fulfil The obligation to fulfil can be categorised into the obligations to facilitate, promote and provide. The obligation to facilitate requires the States to take positive measures to assist the environment to enjoy the right to water. For example, Agenda 21 requires the States to maintain "ecosystem integrity, according to a management principle of preserving aquatic ecosystems, ..., and of effectively protecting them from degradation on a drainage basin basis", "[t]o adopt an integrated approach to environmentally sustainable management of water resources, including the protection of aquatic ecosystems and freshwater living resources".[138] Under the Ramsar Convention 1971, the Contracting Parties "shall designate suitable wetlands within its territory", and "[e]ach Contracting Party shall designate at least one wetland to be included in the List".[139]

The obligation to promote obliges the States to take steps to ensure that there is appropriate education concerning the hygienic use of water, protection of water sources and methods to minimize water wastage. Agenda 21 requires the States to establish national and regional technical and engineering courses on the subjects of water-quality protection and control at existing schools, and education/training courses on water resources protection and conservation.[140] The States are also obliged to fulfil (provide) the right when there is no minimum water to meet the environmental needs for water. In accordance with Agenda 21, the States could implement the following activities, "[r]ehabilitation of important, but degraded, catchment areas, particularly on small islands",

---

136 See para 18.8, Agenda 21. See also arts. 4 and 6, *Agreement on the Cooperation for the Sustainable Development of the Mekong River Basin 1995*; para 6 of the Preamble, art. 3 (3), African Convention on the Conservation of Nature and Natural Resources 2003.

137 See, para. 18.40 (b) (iii), Agenda 21. See also art. 7, *Agreement on the Cooperation for the Sustainable Development of the Mekong River Basin 1995*; art.7, African Convention on the Conservation of Nature and Natural Resources 2003.

138 *Id.*, paras. 18.38(a) and 18.39(g).

139 See art. 2(1) and (4), Ramsar Convention.

140 See para. 18.45, Agenda 21.

"[r]ehabilitation of polluted and degraded water bodies to restore aquatic habitats and ecosystems".[141]

In brief, the obligation to fulfil requires the States to adopt the necessary measures directed towards the reasonable realisation of the environmental right to water. This obligation includes, *inter alia*, according sufficient recognition of this right within the national political and legal systems, preferably by way of legislative implementation; adopting a national water strategy and plan of action to realise this right; ensuring that water is reserved for the environment; and facilitating improved and sustainable access to water.

4.3.3.3 Core obligations

The States should have a core obligation to ensure the satisfaction of, at the very least, minimum essential levels of the environmental right to water, and a number of core obligations in relation to the environmental right to water can be identified, which are of immediate effect: (a) to ensure the minimum essential amount of water that is sufficient and safe for the important parts of the environment;[142] (b) to adopt and implement a national water strategy and plan of action addressing the environment, which should be devised, and periodically reviewed, on the basis of a participatory and transparent process.[143]

4.3.4 Implementation at national level

In accordance with Agenda 21, "[a]ll States, according to their capacity and available resources, and through bilateral or multilateral cooperation, including the United Nations and other relevant organisations as appropriate, could implement" the environmental right to water.[144] That is to say, every State has a margin of discretion in assessing which measures are most suitable to meet its specific circumstances. However, it is clear that each State is under a duty to take whatever steps are necessary or appropriate to ensure that the environment enjoys the right to water. Any national measures designed to realise the

---

[141] *Id.*, para. 18.40 (a) (iv), (e) (i).

[142] For example, see art. 2 (1), Ramsar Convention; it provides that "[e]ach Contracting Party shall designate suitable wetlands within its territory for inclusion in a List of Wetlands of International Importance". See also art. 7(s), UN Convention on Biological Diversity, 1992.

[143] See art. 3(1), Ramsar Convention; it provides that, "[t]he Contracting Parties shall formulate and implement their planning so as to promote the conservation of the wetlands included in the List". See also art. 6, UN Convention on Biological Diversity, 1992; paras. 18.12 (o) (ii), and 18.22, Agenda 21.

[144] See paras. 18.11, 18.12(b), (d), (f), (j), (k) and (l), 18.27, 18.40(A), (B), (D), (E), (F) and (G), 18.50(A), 1859(A), 18.76(C), (D) and (E), and 18.85(a), Agenda 21.

environmental right to water should be consistent with the principle of sustainable development. In accordance with the relevant international documents, the following measures should be adopted in the implementation of the environmental right to water.

4.3.4.1 Formulation of legislation, strategies and policies

In order to implement the environmental right to water, legislation, strategies and policies setting targets and standards and promoting the integrated water resources management should be introduced. Wherever there are existing legislation, strategies and policies, they should be reviewed to ensure that they are compatible with the obligations arising from the environmental right to water, and should be repealed, amended or changed if inconsistent with the environmental right to water.

Principle 11 of Rio Declaration on Environment and Development requires that:

> "States shall enact effective environmental legislation. Environmental standards, management objectives and priorities should reflect the environmental and developmental context to which they apply. Standards applied by some countries may be inappropriate and of unwarranted economic and social cost to other countries, in particular developing countries."[145]

It is clear that the duty to take steps clearly imposes on the States an obligation to adopt a national strategy or plan of action to realise the environmental right to water. And this principle has been reflected in many other international documents. For example, under Article 6 of the UN Convention on Biological Diversity 1992, in accordance with its particular conditions and capabilities, the States shall,

> "(a) Develop national strategies, plans or programmes for the conservation and sustainable use of biological diversity or adapt for this purpose existing strategies, plans or programmes which shall reflect, *inter alia*, the measures set out in this Convention relevant to the Contracting Party concerned; and
>
> (b) Integrate, as far as possible and as appropriate, the conservation and sustainable use of biological diversity into relevant sectoral or cross-sectoral plans, programmes and policies."[146]

The African Convention on the Conservation of Nature and Natural Resources 1968 requires the contracting States to "establish policies for conservation, utilization and development of underground and surface water, ... taking appropriate measures with due regard to – (a) the study of water cycles and the investigation of each catchment area; (b) the co-ordination and planning of water resources development projects; (c) the administration and control of

---

145 See Principle 11, Rio Declaration on Environment and Development.

146 See art. 6, UN Convention on Biological Diversity, 1992.

all water utilization; and (d) prevention and control of water pollution".[147] The African Convention on the Conservation of Nature and Natural Resources 2003 makes a similar provision, that is article 7(2).[148]

It is well known that the human beings have not fully understood the environment, and that the environment is vulnerable to some extent, while there are many other factors that should be taken into consideration. Therefore, it is impossible for the States to adopt detailed legislation. However, the author believes that, the legislation should include: (a) targets or goals to be attained and the time-frame for their achievement; (b) the means by which the purpose could be achieved; (c) the intended collaboration with civil society, private sector and international organisations; (d) institutional responsibility for the process; (e) national mechanisms for its monitoring; and (f) remedies and recourse procedures. Particular measures with respect to minimum flow, site protection, water source protection, species protection etc., should be adopted where appropriate by the States.

In accordance with Agenda 21, the relevant strategy must: (a) be based upon international law and principles relating to the environmental right to water;[149] (b) cover all aspects of the environmental right to water and the corresponding obligations of the States;[150] (c) define clear objectives;[151] (d) set targets or

---

[147] See art. 5(1), African Convention on the Conservation of Nature and Natural Resources 1968.

[148] See art. 7(2), African Convention on the Conservation of Nature and Natural Resources 2003. It reads as follows:

> "The Parties shall establish and implement policies for the planning, conservation, management, utilization and development of underground and surface water, as well as the harvesting and use of rain water, and shall endeavour to guarantee for their populations a sufficient and continuous supply of suitable water, taking appropriate measures with due regard to:
>
> a) the study of water cycles and the investigation of each catchment area,
>
> b) the integrated management of water resources,
>
> c) the conservation of forested and other catchment areas and the co-ordination and planning of water resources development projects,
>
> d) the inventory and management of all water resources, including the administration and control of all water utilization, and
>
> e) the prevention and control of water pollution through, inter alia, the establishment of effluent and water quality standards."

[149] See paras, 15.3 and 18.35, Agenda 21.

[150] *Id.*, para 18.2.

[151] *Id.*, para 18.8.

goals to be achieved and the time-frame for their achievement;[152] and (e) formulate adequate policies and corresponding benchmarks and indicators.[153] The strategy should also establish institutional responsibility for the process;[154] identify the resources available to attain the objectives, targets and goals;[155] allocate resources appropriately according to institutional responsibility;[156] and establish accountability mechanisms to ensure the implementation of the strategy.[157]

Integrated water resources management requires the States to take steps to ensure there is sufficient coordination between the national ministries, and the regional and local authorities in order to reconcile water-related policies.[158] Where implementation of the environmental right to water has been delegated to regional or local authorities, the State still retains the responsibility to comply with its obligations, and therefore should ensure that these authorities have at their disposal sufficient resources to maintain and extend the necessary water services and facilities.

4.3.4.2 Indicators and benchmarks

In order to realise the environmental right to water successfully and assess its implementation, indicators and benchmarks should be established by every State. Firstly, the national strategy plan and action plan should include methods, such as the environmental right to water indicators and benchmarks, by which progress can be closely monitored.[159] Secondly, the States should monitor the extent of the realisation, or the non-realisation, of the environmental right to water.[160] It is essential that the implementation of the environmental right to water can be scientifically monitored, because indicators play an important role in this task.

4.4 CONCLUSION

Under international law, the human right to water as well as the environmental right to water are well discussed in some detail. Although most of the

---

[152] *Id.*, para 18.38.
[153] *Id.*, para 18.11.
[154] *Id.*, para 18.21.
[155] *Id.*, para 18.23.
[156] *Id.*, para 18.39.
[157] *Id.*, para 18.40.
[158] *Id.*, para 18.80.
[159] See art. 3 (2), Ramsar Convention. See also paras. 18.22 , 18.39 (d), 18.43, 18.50 (D), Agenda 21.
[160] See art. 7(b), UN Convention on Biological Diversity, 1992. See also paras. 18.35 18.40 (D) (vi), (F) (i) (G), and 18.85(a), Agenda 21.

international documents employed in this chapter to discuss the details of the human right to water, and, in particular, the environmental right to water, are not legally binding in their forms, all States, including China, are *de facto* under an obligation or a duty to adopt appropriate measures to implement them at national level, *e.g.*, the formulation of legislation, strategies and/or policies, the indicators and benchmarks, and the remedies and accountability, regarding the human right to water and the environmental right to water. These obligations, therefore, will affect the legal forms and substantive content of water rights in China.

In Chapter 5, the legal regimes relating to water rights in selected jurisdictions will be examined, with the aim of discovering experiences that may prove instructive for water-shortage jurisdictions, particularly China.

5

Water Rights Systems in South Africa and South Australia

This chapter will look at the solutions provided in the domestic laws of selected jurisdictions to the similar problems in China as identified in Chapter 3. After the criteria for the selection has been introduced, the analytical approach developed in Chapter 2 will be employed to examine the measures adopted and/or implemented in South Africa and South Australia. This exercise aims to identify any lessons that may prove useful for a water-shortage jurisdiction, in particular China, to improve its legal regime governing its water resources, especially, water rights.

5.1 JUSTIFICATION OF SELECTION

As discussed in Chapter 3, China, the most populous country in the world, faces a serious water crisis. With a population of about 1.3 billion and a land territory of 9.6 million km^2, within which the climate ranges from humid to arid, and different types of water resources (*e.g.*, physical water itself, navigation

ISBN: 1843391082. Published by IWA Publishing, London, UK.

potential, hydropower potential, watercourse, groundwater, surface water, etc.) exist, there is a general water shortage in the whole country, where some areas suffer crucial water shortages and some others are relatively abundant in water resources, and inter-basin water diversion projects exist and should be developed further.[1] Moreover, water pollution and water disasters are also among the primary problems for water resources management in China.[2] As to the legal framework for water resources management, China has a centralised legal regime, although local governments are empowered to adopt measures appropriate to their special regional situations in order to implement the law.[3]

Needless to say, it is difficult to find a jurisdiction that has an identical situation to China concerning water resources. However, under the influence of international law, many jurisdictions worldwide have reformed their water resources management as well as legal regimes respectively to some extent, and some of them have even introduced a complete reform and gained great achievements in some areas. There are also jurisdictions that face a more severe shortage of water than China. By and large, it is possible to select jurisdictions that share similar characteristics in certain aspects, and within which water law has been modernised. Thus, some jurisdictions may be selected for the purpose of comparison, and some aspects relating to water rights in their legal regimes may be employed for comparison. Further, some lessons may be learnt with a view to resolving the relevant problems concerning water rights in water-shortage jurisdictions including China.

As to the criteria for selection of the domestic comparative case studies, the author has used the following factors in a jurisdiction being taken into consideration: (a) a severe shortage of water exists, and the situation is more serious than in China, (b) the inter-basin water transfer exists and is necessary, (c) the water law has been reformed or modernized in written law in the past decade in accordance with modern international law, and (d) relevant achievement(s) in water rights areas has/have been attained.

The reasons for the choice of the above factors are: (a) to fully understand how a water rights system is designed and operates in a more severe water shortage situation where inter-basin water transfer necessary; (b) a modernised water law suggests it may have been influenced by or consistent with the current rules of international law, especially human rights and environmental rights; (c) a written law for water resources may be more convenient for China to learn from, due to its civil law tradition; and (d) the achievement means that the water

---

[1] See the detailed discussion in 3.1.1 and 3.1.2 of Chapter 3.

[2] See the detailed discussion in 3.1.2 of Chapter 3.

[3] See the detailed discussion in 3.2.2 of Chapter 3.

rights system apparently works well and thus can be considered efficient and practical.

South Africa and South Australia fulfil these criteria, and are selected as case studies for the following reasons.[4] (a) As to the total natural renewable water resources per capita received per year, 1,168.2 m^3 in South Africa, and 1,737 m^3 in South Australia, both are lower than the 2,200 m^3 in China. (b) In both jurisdictions, inter-basin water transfer is necessary and has been implemented, while in China, not only inter-basin water transfer exists but also a South-to-North Water Diversion Project, involving many basins, is under construction. (c) South Africa began its water law reform in 1994 after the first real democratic election. Under the present water law in South Africa with the National Water Act 1998 as its core, only the water required to meet basic human needs and maintain environmental sustainability is guaranteed as a right, and the riparian system of allocation has effectively been abolished. In other words, the water law regime in South Africa is relatively new. Australia has committed to many international legal documents relevant to sustainable development, and correspondingly has adopted measures in domestic law.[5] In order to implement the Water Reform Framework 1994 and, subsequently, the Future Information on the National Water Initiative 2004, South Australia has completely reformed its water resources law, and the Water Resources Act 1997 and Natural Resources Management Act 2004 were enacted chronologically.[6]

---

[4] As to the statements relating to China, see the relevant discussion in Chapter 3; regarding those concerning South Africa and Australia, in particular South Australia, see the relevant discussion in the following two parts of this chapter for detail, including necessary sources not given in this paragraph.

[5] See Department for Water Resources of South Australia (DWR), *State Water Plan 2000*, Vol. 1, pp. 3-4.

[6] The Water resources Act 1997, South Australia, assented on 4th October 1997 and entered into force on 2nd July 1997. It has been amended many times; the research of this work is conducted on the version of Reprint No. 10, dated 24th October 2003, available at the Department of Water, Land and Biodiversity Conservation website http://www.dwlbc.sa.gov.au (last visited 31/05/2005).

On 5th August 2004, the Natural Resources Management Act 2004, South Australia, was adopted, and the Water Resources Act 1997 will be repealed by in accordance with Schedule 4 (cl43(1)(c)) of the Natural Resources Management Act 2004. Natural Resources Management Act 2004, available at South Australia's Centre for Natural Resource Management webpage http://www.parliament.sa.gov.au/Catalog/legislation/Acts/N/2004.34.un.htm (last visited 31/05/2005). Some parts of the Natural Resources Management Act 2004, entered into to force on 2nd September 2004. However, Chapter 7 "Management and protection of water resources", which is the key part to repeal the Water Resources Act 1997, was not yet in force by 31 March 2005, the closing date of the research. Undoubtedly, the full implementation of the Natural Resources Management Act 2004 will affect the findings of this research positively, because a more integrated approach is employed in the Natural Resources Act 2004 for South Australia to manage the natural resources,

Under the legal framework of water resources established by the Water Resources Act 1997 and improved by the Natural Resources Management Act 2004, the right to take water from a natural source is well defined and the water rights trading mechanism has been developed on the one hand, while on the other hand, the mechanism for the environment for water is well established.

5.2 LEGAL REGIME IN SOUTH AFRICA

After a brief introduction to South Africa, especially its water law and policy, with a literature review, this section will analyse three aspects of South Africa water law concerning water rights, *i.e.*, water resources and water use, the human right to water and the environmental right to water based on relevant laws and other official documents. The property right on water resources is not discussed in this part, because there are few provisions concerning the property right on water resources in South Africa water law, and therefore, it is of limited value to the Chinese context.

5.2.1 General introduction

In terms of water quantity, South Africa suffers a severe water shortage. Being located in a predominantly semi-arid part of the world, its climate ranges from desert and semi-desert in the west to sub-humid along the eastern coastal area.[7] Its average annual rainfall has decreased by at least 10 percent in the past decade, with an annual rainfall in 1994 of 500 mm, falling to a 450 mm in 2001; both of these figures is only around 50 percent of the world average of 860 mm per year.[8] Lacking any large or navigable rivers, the total flow of all the rivers combined amounted to about 53,500 million m^3 in 1994,[9] but only about 49,000 million m^3 in 2004.[10] At the same time, the rainfall is distributed quite unevenly, *i.e.*, 65 percent of the country gets less than 500 mm per year, and a further 21 percent of the country receives only 200 mm per year.[11] Moreover, there is high average evaporation over most of the country, ranging from about

---

including water resources. Therefore, this work employs the Natural Resources Management Act 2004 for analysis and comparison.

7 See Department of Water Affairs and Forest (DWAF), *White Paper on Water Supply and Sanitation Policy* (November 1994), p. 30; also available at DWAF webpage http://www.dwaf.gov.za/Documents/Policies/WSSP.pdf (last visited 31/05/2005).

8 *Id.* See also DWAF, *infra* note 10, p 15.

9 See DWAF, *supra* note 7.

10 See DWAF, *National Water Resources Strategy (First Edition)* (September 2004), p. 15, 20.

11 See DWAF, *supra* note 7.

1,100 mm per year in the east to over 3,000 mm per year in the west.[12] This situation increases when there is less rainfall. It is quite clear that the evaporation is well in excess of the annual rainfall, and therefore greatly reduces the surface runoff. "Owing to the variability and the high evaporation losses from dams, only about 62 percent or 33,000 million m^3 of the average runoff can be utilised cost-effectively under the current technology".[13] Regarding the groundwater, the geology of South Africa makes it difficult for groundwater to be used. On the one hand, most of the country is underpinned by hard rock formations, and only about 5,400 million m^3 of water may be obtained from underground resources per year; on the other hand, the groundwater is also saline for large areas of the country.[14] According to the White Paper on a National Water Policy for South Africa (April 1997) issued by the Department of Water Affairs and Forest (DWAF), there was only just over 1,200 m^3 of available freshwater per capita per year for a population of 42 million.[15] Data from the World Bank Group suggests that the water resource per capita was only 1,168.2 m^3 in 2000.[16]

The detailed analysis by the DWAF in 2000 shows that, of the total of nineteen water management areas (WMAs), ten were facing a water deficit.[17] Particularly in the northern part of the country, *i.e.*, the WMAs 1-5 and 8-10, surface water, as well as groundwater, is almost completely developed and utilised, and some overdrawing has occurred in localised regions, with little unexploited resource potential remaining.[18] Given this situation, large inter-basin schemes have been constructed, such as the Orange-Fish, Tugela-Vaal, Riversonderend-Berg River and Usutu-Vaal projects.[19] Therefore, in consideration of water allocation, inter-basin water transfers must be taken into account, and special planning requirements and implementation procedures have to be met.[20] Figure 5.1 shows the location of WMR and inter-WMR transfers.

Before the first democratic election in 1994, the water law in South Africa, in its economic content, had evolved over 300 years, mostly concerning the welfare of agricultural land owners initially, and later accommodating

---

12 *Id.*

13 *Id.*

14 *Id.*, pp. 30-31.

15 See DWAF, *White Paper on a National Water Policy for South Africa* (April 1997), p. 9; also available at DWAF webpage http://www.dwaf.gov.za/Documents/Policies/nwpwp.pdf (last visited 31/05/2005).

16 The World Bank Group Data & Statistics, *South Africa Data Profile*, available at http://devdata.worldbank.org/external/CPProfile.asp?SelectedCountry=ZAF&CCODE=ZAF&CNAME=South+Africa&DTYPE=CP (last visited 31/05/2005).

17 See DWAF, *supra* note 10, p. 3.

18 *Id.*, p. 20.

19 See DWAF, *supra* note 7, pp. 4-5.

20 See *supra* note 15, p. 25.

afterwards for the interests of industry and urban cities; as to its form, the system of riparian rights was developed by the courts through combining the Roman-Dutch, English and American laws introduced by the colonialists.[21] Under this riparian rights doctrine, water was divided into public water and private water, with the former being subdivided into normal and surplus water.[22] Under this approach, normal water was subject to proportional distribution, and surplus water could be stored and beneficially used by a riparian owner as far as he was capable of so doing.[23]

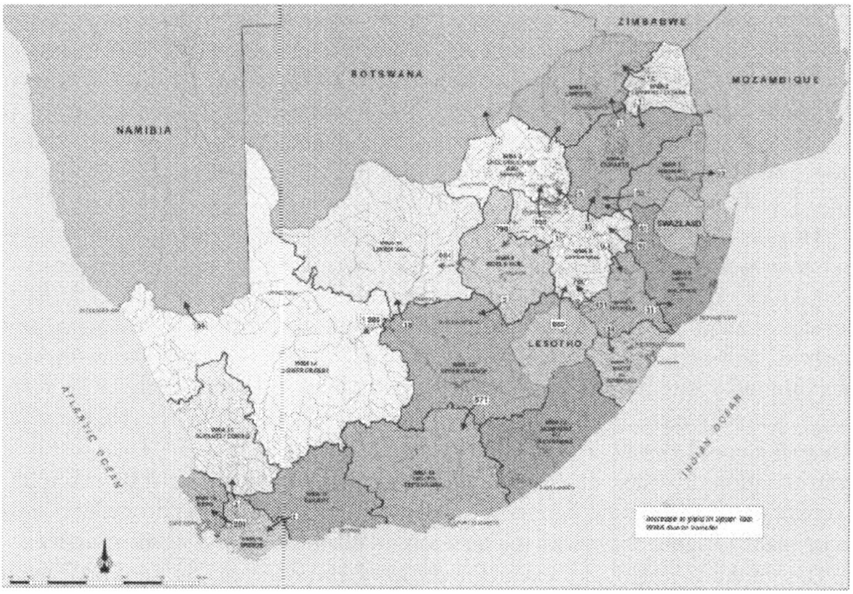

Figure 5.1. Location of WMR and inter-WMR transfers in South Africa (DWAF, National Water Resources Strategy (First Edition), p. 17).

Subsequently, the ever-increasing urbanisation and industrialisation required a re-evaluation of the riparian rights doctrine, and the result was the Water Act of 1956, which carried out incomplete reforms of the riparian rights doctrine.[24] In particular, it produced the following results: (a) the suspension of the principle of riparian rights doctrine, (b) the establishment of certain limitations

---

[21] See DWAF, *supra* note 7, p. 31; also DWAF, *supra* note 15, p. 14.
[22] *Id.*
[23] *Id.*
[24] *Id.*

on the quantity of surplus water stored or diverted by riparian, (c) the establishment of a permit system to manage the use of public water for industrial and urban purposes, (d) the introduction of Government Water Control Areas within which the government could overrule riparian distributions, (e) the launch of Water Boards, which were primarily responsible for ensuring the water supply to urban and industrial consumers, and (f) the normal classification of groundwater as private water, but with restrictions relating to a Subterranean Water Control Area.[25] In conclusion, the legal regime controlling water use in South Africa before the 1994 democratic election was based on the following two important ideas: firstly, "a link between the right to use water and the ownership of land adjacent to the water (the riparian principle)", and, secondly, "a separation between private and public water".[26]

The climates of the regions, either where the Dutch-Roman water law initiated or where English riparian rights doctrine evolved, are quite different from the climate of South Africa.[27] Despite this, the water law in South Africa was based on these water law systems, as discussed above. This foundation, coupled with the apartheid political and economic system, left many people in South Africa with no right or chance to access water to their meet basic human needs.[28] It was estimated that over12 million people did not have access to an adequate supply of potable water and nearly 21 million people lacked basic sanitation.[29] This situation also led to an unsustainable use of water resources, *i.e.*, doing harm to the environment or ecosystem before the first democratic election in 1994.[30]

Since 1st July 1994, on which date the new democratic government was established, the new government has endeavoured to establish a new water law system. The new water law system, which consists of the relevant provisions of Constitution, acts, and policies, has been recently established. Under this system, the central government, as well as the local government, is under a duty to manage water resources in accordance with their respective responsibilities. At the central government level, it is the DWAF that is in charge of managing water resources. In this water rights system, the following documents are of most importance:

- The White Paper on Reconstruction and Development (November 1994);[31]

---

25 *Id.*
26 See DWAF, *supra* note 15, p. 14.
27 See DWAF, *supra* note 7, pp. 31-32.
28 *Id.*, p. 31.
29 *Id.*, p. 1.
30 *Id.*, p. 32
31 Parliament of the Republic of South Africa, *The White Paper on Reconstruction and Development* (November 1994), Notice No. 1954 of 1994, Republic of South Africa, Government Gazette (No. 16085) (23 November 1994); also available at South Africa

- The Constitution of the Republic of South Africa (1996);[32]
- The White Paper on Water Supply and Sanitation Policy (November 1994);
- The Fundamental Principles and Objectives for a New Water Law in South Africa (1996);[33]
- The White Paper on a National Water Policy for South Africa (April 1997);
- The Water Services Act 1997;[34]
- The National Water Act 1998;[35]
- The Water Conservation and Demand Management National Strategy Framework (Draft May 1999);[36]
- The National Water Resources Strategy (First Edition) (September 2004).

Under the new water law regime in South Africa, the two aspects of water rights, *i.e.*, the human right to water and the environmental right to water are emphasized. After several key concepts have been introduced, these two aspects will now be discussed in detail.

---

government webpage http://www.gov.za/gazette/whitepaper/1994/16085.pdf (last visited 31/05/2005).

[32] The Constitution of the Republic of South Africa (1996), Act No. 108 of 1996, adopted on 8 May 1996 by the Constitutional Assembly, approved by the Constitutional Court on 4 December 1996, took effect on 4 February 1997, and amended on 11 October 1996, 19 June 2002, 19 March 2003 and 11 April 2003. Its present version available at South Africa government webpage http://www.gov.za/constitution/1996/96cons.htm (last visited 31/05/2005).

[33] DWAF(1996), *Fundamental Principles and Objectives for a New Water Law in South Africa*, approved by the Cabinet in November 1996, as Appendix 1 of White Paper on a National Water Policy for South Africa (April 1997).

[34] Water Services Act 1997, Act No. 108 of 1997, assented to 27 November 1997, Republic of South Africa, Government Gazette (No. 18522) (19 December 1997); also available at DWAF webpage http://www.dwaf.gov.za/Documents/Legislature/wsa97.pdf (last visited 31/05/2005).

[35] National Water Act 1998, Act No. 36 of 1998, assented to 20 August 1998, Republic of South Africa, Government Gazette (No. 19182) (26 August 1998); also available at DWAF webpage http://www.dwaf.gov.za/Documents/Legislature/nw_act/NWA.pdf (last visited 31/05/2005). It was amended on 2 December 1999 by National Water Amendment Act (1999), Act No. 45 of 1999, Republic of South Africa, Government Gazette (No. 20706) (6 December 1999); also available at DWAF webpage http://www.dwaf.gov.za/Documents/Legislature/wsa97.pdf (last visited 31/05/2005).

[36] DWAF, *Water Conservation and Demand Management National Strategy Framework* (Draft May 1999), available at DWAF webpage http://www.dwaf.gov.za/Documents/Policies/draftNWCSF.pdf (last visited 31/05/2005).

5.2.2 Water resource and water use

As mentioned earlier, water law regulates water resources and/or water use.[37] In South Africa, "water resource" and "water use" are under the regulation of water law at the same time. Therefore, it is necessary to define these two concepts in South Africa water law and policy.

5.2.2.1 Water resource

In the White Paper on a National Water Policy for South Africa, "water resource" is defined as "the water in the natural water cycle from which we draw the water required to meet human needs",[38] and "water cycle" is defined as "the natural cycle whereby water falls from clouds onto the land as rain, seeps into 'aquifers' underground or drains into rivers and eventually flows to the sea where much of the evaporation which takes the water up into the clouds again occurs".[39] Among the key proposals that guide water management in South Africa, there is one which concerns the understanding of "water resource", *i.e.*, "[a]ll water in the water cycle whether on land, underground or in surface channels, falling on, flowing through or infiltrating between such systems, will be treated as part of the common resource and to the extent required to meet the broad objectives of water resources management, will be subject to common approaches".[40] This idea has been reflected in the National Water Act 1998. In accordance with section 1(1) (xxvii) of the act, "'water resource' includes a watercourse, surface water, estuary, or aquifer". In the National Water Resources Strategy (First Edition), on the basis of the definition in the National Water Act 1998, water resource is explained as "surface water found in watercourses (rivers and streams), impoundments (dams), wetlands and estuaries, and groundwater found in underground aquifers".[41]

It is obvious that, in South Africa, although "water resource" is understood as all water in the water cycle from a policy perspective, it only includes watercourse, surface water, estuary and aquifer in legal forms. In other words, as an ethic for managing water resources, all water in the water cycle should be taken into consideration. However, this geological definition of water resource could not be regulated in legal forms, because the legislation has its own requirements and forms.

---

[37] See the relevant discussion in 2.1 of Chapter 2.

[38] See DWAF, *supra* note 15, p. 37.

[39] *Id.*

[40] *Id.*, p. 3.

[41] See DWAF, *supra* note 10, p. 12.

5.2.2.2 Water use

Under the National Water Act 1998, water use is defined broadly in section 21, which reads as follows:

> "For the purposes of this Act, water use includes –
>
> (a) taking water from a water resource;
>
> (b) storing water;
>
> (c) impeding or diverting the flow of water in a watercourse;
>
> (d) engaging in a stream flow reduction activity contemplated in section 36;
>
> (e) engaging in a controlled activity identified as such in section 37(1) or declared under section 38(1);
>
> (f) discharging waste or water containing waste into a water resource through a pipe, canal, sewer, sea outfall or other conduit;
>
> (g) disposing of waste in a manner which may detrimentally impact on a water resource;
>
> (h) disposing in any manner of water which contains waste from, or which has been heated in, any industrial or power generation process;
>
> (i) altering the bed, banks, course or characteristics of a watercourse;
>
> (j) removing, discharging or disposing of water found underground if it is necessary for the efficient continuation of an activity or for the safety of people; and
>
> (k) using water for recreational purposes." [42]

It is clear that all activities that might affect water resources fall within the scope of water use.

In order to control water use, the Act catalogues three kinds of water use, *i.e.*, (a) water use without a licence; (b) water use authorised by a licence; (c) water use under a licence, permit or other authorisation under any other law than the National Water Act 1998.[43] And any person, who uses water:

> "(a) must use the water subject to any condition of the relevant authorisation for that use;
>
> (b) is subject to any limitation, restriction or prohibition in terms of this Act or any other applicable law;

---

[42] See section 21, National Water Act 1998.

[43] *Id.*, section 22 (1) and (3).

(c) in the case of the discharge or disposal of waste or water containing waste contemplated in section 21(f), (g), (h) or (j), must comply with any applicable waste standards or management practices prescribed under section 26(1)(h) and (i), unless the conditions of the relevant authorisation provide otherwise;

(d) may not waste that water; and

(e) must return any seepage, run-off or water containing waste which emanates from that use, to the water resource from which the water was taken, unless the responsible authority directs otherwise or the relevant authorisation provides otherwise." [44]

In other aspects, all water use is not permanent and is granted for a reasonable period;[45] under strict provisions stipulated by the Act, water use allocation could be transferred or traded between users with the consent of the Minister.[46]

In brief, all water use affecting water resources in South Africa are subject to the regulation of water law, its categories, conditions and transfers, as well as its durations.

5.2.3 The human right to water

Under the White Paper on Reconstruction and Development, one of the five key programmes of the Reconstruction and Development Programme is to meet basic needs. As to the content of the basic needs of the people in South Africa, this should "extend from job creation, land and agrarian reform to … water and sanitation, …".[47] In Chapter 2 "Bill of Rights", an indispensable part of the Constitution of the Republic of South Africa, there is one provision particularly relevant to the human right to water, and it is section 27. The substantive content of the provision relating to the human right to water reads as follows:

"(1) Everyone has the right to have access to …

b. sufficient food and water; and …

(2) The state must take reasonable legislative and other measures, within its available resources, to achieve the progressive realisation of each of these rights."[48]

From the above stipulation, it is very clear that the right to water is one of human rights in South Africa. In order to realise the human right to water stipulated in the Constitution, relevant acts and policies have been adopted, which will be discussed below.

---

[44] *Id.*, section 22 (2).

[45] *Id.*, section 28.

[46] *Id.*, section 25.

[47] See Parliament of the Republic of South Africa, *supra* note 31, sections 1.4.1 and 1.4.3, p. 9.

[48] See section 27, the Constitution of the Republic of South Africa.

5.2.3.1 The water required to meet basic human needs as a human right

In the processes of enacting a new water service act and a new national water act, the Fundamental Principles and Objectives for a New Water Law in South Africa was developed by the DWAF and was approved by the Cabinet in November 1996.[49] In this document, 28 principles are introduced.[50] Of these, Principles 8, 10 and 13 relate directly to the water required to meet basic human needs as a human right. Principle 8 declares that, "[t]he water required to ensure that all people have access to sufficient water shall be reserved".[51] Principle 10 further provides that, "[t]he water required to meet the basic human needs referred to in Principle 8 and the needs of the environment shall be identified as `The Reserve' and shall enjoy priority of use by right. The use of water for all other purposes shall be subject to authorisation".[52] Further, Principle 13 establishes the responsibility of the government to ensure and promote the realisation of the human right to water. It provides that:

> "As custodian of the Nation's water resources, the National Government shall ensure that the development, apportionment, management and use of those resources is carried out using the criteria of public interest, sustainability, equity and efficiency of use in a manner which reflects its public trust obligations and the value of water to society while ensuring that basic domestic needs, the requirements of the environment and international obligations are met." [53]

In the National Water Act 1998 as well as the White Paper on a National Water Policy for South Africa, these principles are all fully reflected. Under the White Paper on a National Water Policy for South Africa, several key proposals are set forth, which are employed to guide water management in South Africa.[54] Of these, two are of vital importance to the human right to water. The first is that "[o]nly that water required to meet basic human need and maintain environment sustainability will be guaranteed as a right. This will be known as Reserve".[55] And the second is that "to promote equitable access to water for basic human needs, provision will also be made for some or all of these charges to be waived".[56]

---

[49] See DWAF, *supra* note 15, p. 6.
[50] See DWAF, *supra* note 33.
[51] *Id.*, Principle 8.
[52] *Id.*, Principle 10.
[53] *Id.*, Principle 13.
[54] See DWAF, *supra* note 15, pp. 3-4.
[55] *Id.*, p. 4.
[56] *Id.*

Regarding the National Water Act 1998, in section 2, within which the purposes of the Act are identified, "meeting the basic human needs of present and future generation" is listed as the primary purpose.[57] In order to realise this purpose, the following systems and/or provisions have been implemented.

The first one is the reserve system. In accordance with section 1(1) (xviii) of the National Water Act 1998, "[r]eserve" means the quantity and quality of water required to satisfy basic human needs and to protect the aquatic ecosystem.[58] In detail, the legal content for satisfying basic human needs means that, the quantity and quality of water required "to satisfy basic human needs by securing a basic water supply, as prescribed under the Water Services Act 1997, for people now or who will, in the reasonably near future, be (i) relying upon; (ii) taking water from; or (iii) being supplied from, the relevant water resources".[59]

The second one is the water management strategy system. The National Water Act 1998 regulates the substantive content of the national water resource strategy in section 6 (1), which provides that the national water resource strategy must:

> "(a) set out the strategies, objectives, plans, guidelines and procedures of the Minister and institutional arrangements relating to the protection, use, development conservation, management and control of water resources within the framework of existing relevant government policy in order to achieve –
>
> (i) the purpose of this Act; and...
>
> (b) provide for at least –
>
> (i) the requirements of the Reserve and identify ,where appropriate, water resources from which particular requirements must be met;
>
> ..."[60]

Under the National Water Resources Strategy (First Edition), a proportion of the available water is under the direct control of the Minister of DWAF, and the Reserve is given the highest priority, in which the first objective is to ensure that sufficient quantities of water of appropriate quality are reliably available to provide for basic human needs.[61] Further, in order to accomplish this purpose, a quantity of 25 litres per person per day has been incorporated under the urban and rural requirements in the present policy.[62]

The third one is the reasonable domestic water use without a license. In accordance with sections 22 (1) (a) and 39 of the National Water Act 1998, a

---

[57] See section 2(a), National Water Act 1998.
[58] *Id.*, section 1(1) (xviii).
[59] *Id.*
[60] *Id.*, section 6(1) (a) and (b).
[61] See DWAF, *supra* note 10, p. 52.
[62] *Id.*

person may use water without a license under the following three circumstances, *i.e.*, if that water use is permissible, i) under Schedule 1; ii) as a continuation of an existing lawful use; or iii) in terms of a general authorisation, subject to Schedule 1, issued under section 39.[63] Under section (1) (a) and (b) (i) of Schedule 1, a person may "take water for reasonable domestic use in that person's household, directly from any water resource to which that person has lawful access", or "take water for use on land owned or occupied by that person, for (i) reasonable domestic use".[64]

5.2.3.2 The right of access to basic water services

Needless to say, in modern society with serious water pollution, especially in urban areas, it is almost impossible for one to obtain safe or drinkable water directly from natural water resources in most cases. Therefore, without access to basic water services, it is almost impossible for a person to fulfil his/her right to obtain water to meet his/her basic human needs. Therefore, to some extent, accessing basic water services is the precondition for obtaining water to meet basic human needs. In the White Paper on Water Supply and Sanitation Policy, eight Policy principles are introduced in Section C, and "Basic Services are a human right" is among them.[65] In the Fundamental Principles and Objectives for a New Water Law in South Africa, Principle 25 states that, "[t]he right of all citizens to have access to basic water services (the provision of potable water supply and the removal and disposal of human excreta and waste water) necessary to afford them a healthy environment on an equitable and economically and environmentally sustainable basis shall be supported".[66] These principles have been adopted in the Water Services Act 1997, as will be discussed below.

In the Water Services Act 1997, "the right of access to basic water supply and basic sanitation necessary to ensure sufficient water and an environment not harmful to health or well-being" is recognised, and that "there is a duty on all spheres of Government to ensure that water supply services and sanitation services are provided in a manner which is efficient, equitable and sustainable" is acknowledged.[67] According to section 2 (a), the first of the main objects of the Water Services Act 1997 is to provide for "the right of access to basic water supply and the right to basic sanitation necessary to secure sufficient water and

---

[63] See sections 22 (1) (a) and 39, National Water Act 1998.

[64] *Id.*, Section (1) (a) and (b) (i) of Schedule 1.

[65] DWAF, *supra* note 7, p. 8.

[66] See DWAF, *supra* note 15, p. 36.

[67] See paras. 1-2 of the Preamble, Water Services Act 1997.

an environment not harmful to human health or well-being".[68] In section 1, "basic water sanitation" and "basic water supply" are defined correspondingly, the former term "means the prescribed minimum standard of services necessary for the safe, hygienic an adequate collection, removal, disposal or purification of human excreta, domestic waste-water and sewage from households, including informal households", while the latter "means the prescribed minimum standard of water supply services necessary for the reliable supply of a sufficient quantity and quality of water to households, including informal households, to support life and personal hygiene". [69] Further, the relevant rights and duties/responsibilities are declared in section 3, which reads as follows:

> "(1) Everyone has a right of access to basic water supply and basic sanitation.
>
> (2) Every water services institution must take reasonable measures to realise these rights.
>
> (3) Every water services authority must, in its water services development plan, provide for measures to realise these rights.
>
> (4) The rights mentioned in this section are subject to the limitations contained in this Act." [70]

Moreover, section 5 further provides that,

> "If the water services provided by a water services institution are unable to meet the requirements of all its existing consumers, it must give preference to the provision of basic water supply and basic sanitation to them." [71]

In order to deal with the problem of poor people not having the chance to obtain basic water, President Thabo Mbeki announced the Free Basic Water Policy in September 2000.[72] In January 2003, the National Cabinet approved a three-stage Free Basic Water Implementation Programme, under which each household is supplied with 6 m^3 (6,000 litres) of safe water per month without charge.[73]

---

[68] *Id.*, section 2 (a).

[69] *Id.*, section 1(ii) and (iii).

[70] *Id.*, section 3.

[71] *Id.*, section 5.

[72] See Mbeki, Thabo (2000) *Address to the COSATU 7th National Congress*, 18 September, 2000, available at http://www.cosatu.org.za/congress/cong2000/tm000918.htm (last visited 31/05/2005).

[73] See DWAF, "*Free Basic Water" Implementation Strategy Document (Version 1)* (May 2001), p. 4; available at DWAF webpage http://www.dwaf.gov.za/FreeBasicWater/docs/Implementation%20Strategy%20version%208.3.pdf (last visited 31/05/2005). This amount is calculated for a household of eight people on the World Health Organization standard of 25 litres per person per day.

To conclude this section, the Water Services Act 1997, which derived from the White Paper on Water Supply and Sanitation Policy, and the National Water Act 1998, which directly originated from the Fundamental Principles and Objectives for a New Water Law in South Africa and the White Paper on a National Water Policy for South Africa, both make the human right to water a priority. As to the substantive content of the human right to water, it is regulated in two aspects: (a) the water required to meet basic human needs enjoys priority of use by right, and (b) each citizen has access to basic water services.

5.2.4 The environmental right to water

The water law of South Africa emphasises the protection of the environment relating to water. As discussed in more detail below, this can be identified in law as well as in policy.

From the Constitution of the Republic of South Africa, one may find the origin of the environmental right to water. It provides that any development and utilisation of natural resources, undoubtedly including water resources, must be environmentally sustainable.[74] In the Fundamental Principles and Objectives for a New Water Law in South Africa, of the 28 principles employed, at least nine are related to the environmental right to water, *i.e.*, Principles 5, 6, 7, 9, 10, 13, 15, 16, and 17.

In Principle 5, the unity of the water cycle and the interdependence of its elements are recognised.[75] Principle 6 states that, "[t]he variable, uneven and unpredictable distribution of water in the water cycle (in South Africa) should be acknowledged";[76] and Principle 7 describes the objective of the government in its management of water as being "to achieve optimum, long term, environmentally sustainable social and economic benefit for society from their use".[77]

Further, Principle 9 emphasises the role of water to the environment, and reads as follows:

> "The quantity, quality and reliability of water required to maintain the ecological functions on which humans depend shall be reserved so that the human use of water does not individually or cumulatively compromise the long term sustainability of aquatic and associated ecosystems."[78]

Moreover, Principle 10 declares that, "[t]he water required to meet the basic human needs referred to in Principle 8 and the needs of the environment shall be

[74] See section 24, The Constitution of the Republic of South Africa.
[75] See DWAF, *supra* note 15, p. 35.
[76] *Id.*
[77] *Id.*
[78] *Id.*

identified as 'The Reserve' and shall enjoy priority of use by right. The use of water for all other purposes shall be subject to authorisation";[79] and Principle 13 set up the responsibility of the government to ensure and promote the realisation of the environmental right to water.[80]

Principle 15 demands that, "[w]ater quality and quantity are interdependent and shall be managed in an integrated manner, which is consistent with broader environmental management approaches";[81] while Principle 16 provides that, "[w]ater quality management options shall include the use of economic incentives and penalties to reduce pollution; and the possibility of irretrievable environmental degradation as a result of pollution shall be prevented."[82]

Principle 17 stipulates that, "[w]ater resource development and supply activities shall be managed in a manner which is consistent with the broader national approaches to environmental management."[83]

In the White Paper on a National Water Policy for South Africa, the above principles are fully reflected, and there are many paragraphs relating to the environmental right to water. Firstly, the document states several key proposals that should be employed to guide water management in South Africa. Of these, two are of great importance to the environmental right to water. The first one is that all water in the water cycle "will be treated as part of the common resource and to the extent required to meet the broad objectives of water resource management, will be subject to common approaches";[84] and the second one is that, "[o]nly that water required to meet basic human need and maintain environment sustainability will be guaranteed as a right. This will be known as Reserve".[85]

Secondly, Section 4.2, which is entitled "Optimum resource use and protection", explains the three important concepts, *i.e.*, "optimum", "environmentally sustainable" and "social economic benefit", and states the interdependent relations between them.[86] As to the meaning of the sustainable use of water resources, it is explained as, "even where the immediate demands for development are very high, society must find different development approaches which make sure that the use of water resources does not destroy the ability to recover".[87]

Thirdly, the public trust obligations of the government and the priorities in water resources management are examined. In Section 5.1.2, in which public

---

79 *Id.*

80 *Id.*, p. 36.

81 *Id.*

82 *Id.*

83 *Id.*

84 See DWAF, *supra* note 15, p. 4.

85 *Id.*

86 *Id.*, pp. 12-13.

87 *Id.*, p. 13.

trust is discussed, the government states that it will carry out its public trust obligations in a way, which "makes sure that the requirements of the environment are met; takes into account the interconnected nature of the water cycle".[88] In Section 5.2, the priorities are discussed in detail, and the environmental requirement for water is among the three types of priorities, *i.e.*, basic human needs, environmental requirements, and international obligations.[89] The document declares that, "[a]fter providing for the basic needs of citizens, the only other water that is provided as a right, is the Environmental Reserve – to protect the ecosystems that underpin our water resources, now and into the future".[90]

Finally, the protection of water resources is discussed in detail in terms of three aspects, *i.e.*, quality and its protection, resources protection, and source directed controls and their enforcement, in Section 6.3.[91]

In the National Water Act 1998, there are some vital provisions that regulate the environmental right to water. The Preamble recognises the sustainable use of water, the protection of the quality of water resources, and the integrated management of water resources.[92] Further, the purposes of the Act include "promoting the efficient, sustainable and beneficial use of water in the pubic interest", "protecting aquatic and associated ecosystems and their biological diversity" and "reducing and preventing pollution and degradation of water resources".[93]

As to the concrete provisions, the environmental right to water is regulated primarily in terms of three aspects, *i.e.*, the Reserve system, the government's duty, power and responsibility, and the role of the public; these will be discussed in further detail below.

The first aspect is the Reserve system. In accordance with the National Water Act 1998, "Reserve" means the quantity and quality of water required to satisfy basic human needs, and "to protect aquatic ecosystem in order to secure ecologically sustainable development and use of the relevant water resource".[94] In Chapter 4, the protection of water resources is regulated. In detail, the determination, preliminary determination, and effect of Reserve are stipulated.[95]

The second aspect is the duty, power and responsibility of the government to protect the water related environment. The government is under a legal

---

88 *Id.*, p. 14.
89 *Id.*, p. 16.
90 *Id.*
91 *Id.*, pp. 18-20.
92 See paras. 4-6 of the Preamble, National Water Act 1998.
93 *Id.*, section 2.
94 *Id.*, section 1(1) (xviii).
95 *Id.*, sections 16-18.

obligation to ensure that water is being protected, used, developed, conserved, managed and controlled in a sustainable manner, and the Minister of DWAF is ultimately responsible for promoting environmental values.[96] Section 6 of the National Water Act 1998 provides that the national water resource strategy must "(a) set out the strategies, objectives, plans, guidelines and procedures of the Minister and institutional arrangements relating to the protection, use, development conservation, management and control of water resources within the framework of existing relevant government policy in order to achieve (i) the purpose of this Act; and … ; (b) provide for at least (i) the requirements of the Reserve and identify ,where appropriate, water resources from which particular requirements must be met".[97] In accordance with section 45 (2), a proposed allocation schedule must reflect the quantity of water to be assigned to the Reserve. On reviewing a licence, of the three situations under which a responsible authority may amend any conditions of the licence, two relate to the environmental right to water, *i.e.*, (a) it is necessary or desirable to prevent deterioration or further deterioration of the quality of the water resource, and (b) there is insufficient water in the water resource to accommodate all authorised water uses after allowing for the Reserve and international obligations.[98] Section 67 provides that, in an urgent situation, or in cases of extreme urgency involving the protection of a water resource or the environment, the Minister of DWAF may adopt measures dispensing with certain requirements of the Act. According to the interpretation in the National Water Resources Strategy, "[t]o achieve sustainable use of water … by … striking a balance between water availability and legitimate water requirements, and by implementing measures to protect water resources" is one of the three fundamental objectives for managing South Africa's water resources.[99] Moreover, a proportion of the available water is under the direct control of the Minister of DWAF, and the Reserve is given the highest priority, within which the provision of water for safeguarding and sustaining healthy ecosystems follows the first objective -- ensuring basic human needs -- to be met.[100] Further, the National Water Resources Strategy (First Edition) discusses the protection of water resources in detail in Part 1 of Chapter 3.[101]

The third aspect is the role of the public, including environmental interest groups, in protecting the environmental right to water under a proper legal procedure. Under the National Water Act 1998, the public is entitled to participate in environmental impact assessment procedures in accordance with

---

96 *Id.*, section 3.
97 *Id.*, section 6 (1) (a) and (b).
98 *Id.*, section 49 (2).
99 See DWAF, *Supra* note 10, p. 7.
100 *Id.*, p. 52.
101 *Id.*, pp. 56-62.

the law.[102] According to section 81, when the members of a governing board of a catchment management agency[103] are appointed, the Minister of DWAF must balance the interests of water users, potential water users, local and provincial government and environmental interest groups. Moreover, in accordance with the National Environmental Management Act 1998, (a) any person or group of persons may in the interest of protecting the environment enforce environmental laws, and (b) any person may in the interest of the protection of the environment institute and conduct a private prosecution.[104]

5.3 LEGAL REGIME IN SOUTH AUSTRALIA

Following a succinct literature review on Australia, especially South Australia's water law and policy, this part analyses three aspects of South Australia water law concerning water rights, *i.e.*, the definition of water resource, the property right on water resources and the environmental right to water based on the relevant laws and other official documents. South Australia does not have a similar apartheid history to South Africa, and the human right to water does not have any particular status like in South Africa, although the human right to water is regulated to some extent.[105] Thus, the human right to water will not feature in the discussion of this part.

5.3.1 General introduction

With an average annual precipitation of less than 600 mm in 80 percent of the continent, accompanied by an uneven distribution in time as well as in region, Australia is the driest inhabited continent on the earth.[106] Moreover, a high

---

[102] See section 100, National Water Act 1998.

[103] Catchment management agency is a kind of responsible authority other than the Minister of DWAF, and a catchment management agency exercises relevant power or duty assigned by the Minister in respect of water uses.

[104] See sections 32-33, National Environmental Management Act 1998, Act No. 107 of 1998, assented to 19 November 1998, Republic of South Africa, Government Gazette (No. 19519) (27 November 1998); also available at South Africa Government webpage http://www.gov.za/acts/1998/a107-98.pdf (last visited 31/05/2005).

[105] For example, in accordance with section 7 (5) and (7) of the Water Resources Act 1997, South Australia, a person who has lawful access to water resources is entitled to take water for domestic purposes from the relevant sources without a licence, even the relevant sources are prescribed.

[106] See [Commonwealth of Australia] Bureau of Meteorology, *Australia - Climate of Our Continent*, available at http://www.bom.gov.au/lam/climate/levelthree/ausclim/zones.htm (last visited 31/05/2005).

evaporation rates leads to a runoff of just 11% and a groundwater recharge of only 1% of average precipitation. This gives Australia approximately 5% of the global land area with only 1% of the world river flows.[107] Several states suffer from a shortage of water resources.[108] As a federal country, Australia consists of six states and two territories, and South Australia forms one of the six states. Under the federal system of government, the states and territories have primary responsibility for managing the natural resources, including the water resources, within their jurisdiction respectively.

Being located in the mid-south of the continent with the Southern Ocean to the south and being bordered on the other sides by land, South Australia covers an area of 985,335 km^2, with a population of 1,534,250 on 30 June 2004, and it is by far the driest of the Australia states and territories.[109] According to the South Australia State Water Plan, the water resources situation may be summarized in more detail as follows.[110]

The annual precipitation of Antarctica estimated is about 157 mm. See Bromwich D.H., et al., *The atmospheric hydrologic cycle over the Southern Ocean and Antarctica from operational numerical analyses*, 123 Monthly Weather Review 3518.

[107] See National Land & Water Resources Audit, *Australian Water Resources Assessment 2000*, available at http://audit.ea.gov.au/ANRA/water/docs/national/Water_Contents.html (last visited 31/05/2005).

[108] For example, many areas in Queensland and New South Wales have not seen rain for the last two years, and some parts have experienced the lowest rainfall since records began in 1900. See, Australia's drought halves grain harvest, available at http://news.bbc.co.uk/1/hi/world/asia-pacific/2372393.stm (last visited 31/05/2005).

[109] Australia Bureau of Statistics, *South Australia at a Glance 2005*.

[110] See DWR, *State Water Plan 2000*, Vol. 1, pp. 13-14; Vol. 2, at 1-4.

The reader should note that, as the first statutory plan relating to water resources in South Australia, the State Water Plan 2000 provides the top level of water resources policy and inventory information. In detail, it gives the policy framework for water resources management and use throughout South Australia, with catchment water management plans, water allocation plans and local water management plans providing the regional and local focus. The Water Resources Act 1997 requires these plans to be in conformity with the State Water Plan and provides for regular reviews to ensure uniformity through a holistic process of community involvement. The Plan consists of two volumes. Further, there is an additional document Explanatory Documents, consisting of papers on four key water resource topics and providing background information, but it does not form part of the Plan. Under The Water Resources Act 1997, the State Water Plan shall be reviewed of after five years, that is to say, the Plan which would be due in 2005. However, the Natural Resources Management Act 2004 requires a State Natural Resources Management Plan to be prepared that will address policy issues, and document the nature and condition of all relevant natural resources. Therefore, the State Water Plan will be adopted as the initial State Natural Resources Management Plan. Due to the validation of the State Water Plan before the adoption of the State Natural Resources Management Plan, this work will conduct the research on the State Water Plan 2000.

Like the whole country, South Australia's water resources are unevenly distributed. With less than 15% of the State total area and the majority of its population, the southern and coastal areas enjoy precipitation in the range of 300 mm to over 1,000 mm per year. Rainfall occurs mainly in winter and early spring. With a high evaporation rate in summer, the rivers always tend to become a series of pools or even dry out. In the rest of the State, the situation is more serious because of the climate varying from semi-arid to arid. Irregular, concentrated or very strong rainfall causes local widespread flooding on the one hand, while, on the other hand, high evaporation rates dry out surface water quickly. As a result, limited surface water resources are produced within the State, and occur predominantly in the Mount Lofty and Flinders Ranges. The State's major surface water resources are in the Murray-Darling and Lake Eyre Basins. However, groundwater is accessible over more than half of the State all year round and consequently is the major natural source of water for much of the State. In total, the amount of reliable water resources is 2,610 million m^3 per year, which consists of 920 million m^3 surface water, 1,440 million m^3 groundwater, and 250 million m^3 runoff and reclaimed water, although the quantity and quality vary from region to region. Of the above water resources, a portion is salty, and therefore, this component of the water resources is of limited use or has to be treated before normal use. It may be calculated that, the total renewable water resources per capita per year in South Australia is only about 1,735 m^3.

Because of the unevenly natural distribution of the water resources, since the end of the Second World War, inter-basin/catchment pipelines have been constructed to transport water and meet the increasing water demand in the newly developed areas (see Figures 5.2 and 5.3).[111]

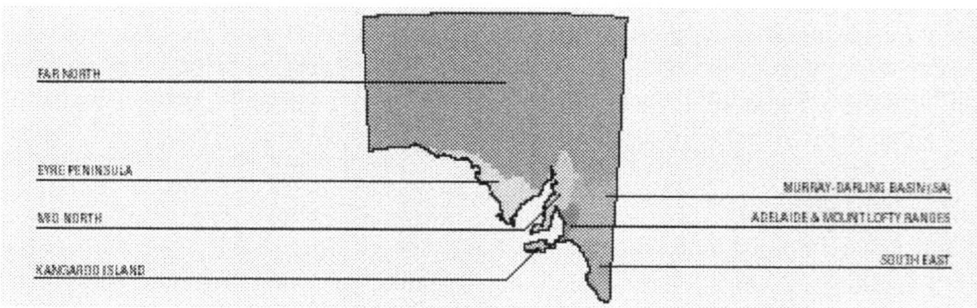

Figure 5.2. South Australia water resources management regions (DWR, State Water Plan 2000, Vol. 1, p. 15).

[111] See DWR, *State Water Plan 2000*, Vol. 2, p. 1.

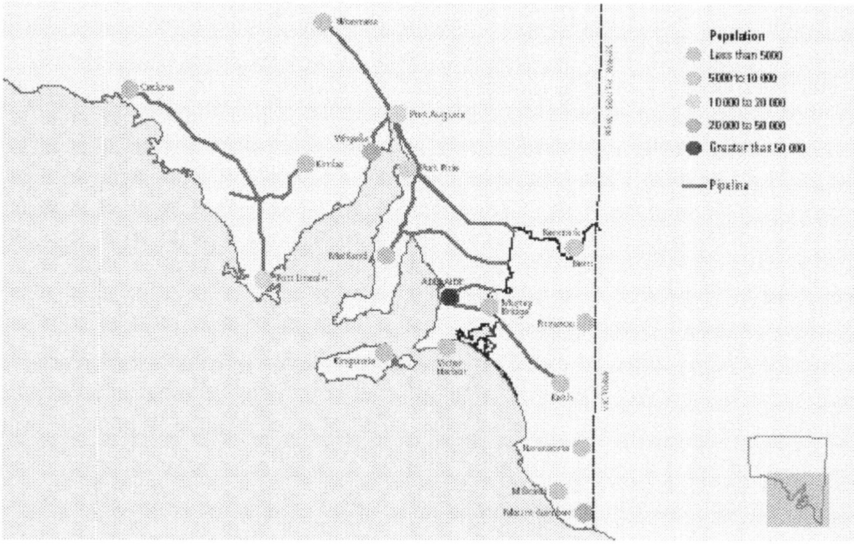

Figure 5.3. South Australia major water supply pipelines and population centres (DWR, State Water Plan 2000, Vol. 2, p. 2).

As to the evolution of water law, this may be divided into two chronological phases, *i.e.*, the common law phase and the statutory phase.[112] The statutory phase began in 1976, in which year the Water Resources Act 1976 was adopted, which gradually modified the common law rules relating to water, and finally abolished them to some extent.[113] In 1997, "[r]ights at common law in relation to the taking of naturally occurring water are abolished".[114] Up to now, all water resources are under the administration of the State through the Minister for Environment and Conservation via the Department of Water, Land and Biodiversity Conservation.

The present water law reform began on 25 February 1994, at which time the COAG endorsed a national key water resource policy, *i.e.*, the Water Reform Framework 1994. In this Framework, the introduction of a comprehensive system of water allocations is recommended, and the reform package involves six elements, *i.e.*, pricing, water allocation or entitlements, trading in water allocation or entitlements, institutional reform, consultation and public

[112] See DWR, *Who owns water?* pp. 8-10, in DWR, *State Water Plan 2000*, Explanatory Documents.

[113] *Ibid.*, pp. 9 – 10.

[114] See section 7 (9), Water Resources Act 1997. This provision is reconfirmed by the section 124 (8) of the Natural Resources Management Act 2004.

education, and the environment.[115] South Australia implemented this reform framework through relevant measures; among which, the enactment of the Water Resources Act 1997 is the most important. As the first important legislation regulating access to and utilisation of water, the Water Resources Act 1997 was assented to on 10 April 1997 and came into force on 2nd July of the same year, and has been amended many times.[116] The legal framework for water resources management in South Australia established by the Act is based on the statutory rights to access water, the power accorded to the Minister, and the hierarchy of water resource management plans (water plans) – The State Water Plan, catchment water management plans, water allocation plans and local water management plans.

On 29 August 2003, in order "to increase the productivity and efficiency of water use, sustain rural and urban communities, and to ensure the health of river and groundwater systems", the COAG agreed to refresh the Water Reform Framework 1994, and a Proposed National Water Initiative was recommended. Later, on 25 June 2004, the COAG approved the Future Information on the National Water Initiative 2004 at a meeting held in Canberra. Following the Initiative, on 5th August 2004, the Natural Resources Management Act 2004, South Australia, was adopted, and the Water Resources Act 1997 will be repealed in accordance with Schedule 4 (cl 43(1)(c)) of the Natural Resources Management Act 2004.

Other acts, such as the Development Act 1993,[117] the Waterworks Act 1932,[118] the Sewerage Act 1929[119] and the Environmental Protection Act 1997,[120] also play an important role in certain aspects to some extent in the

---

115 See COAG, *Water Reform Framework 1994*, sections 3-8.

116 See *Legislative History* of Water Resources Act 1997, available at http://www.austlii.edu.au/au/legis/sa/consol_act/wra1997197.txt (last visited 31/05/2005).

117 The Development Act 1993, South Australia, assented on 27 May 1993 and entered into force on 15 January 1994. It has been amended many times; the version of Reprint No. 15, dated 14 June 2001, available at http://www.parliament.sa.gov.au/lnkdocs/developm.htm (last visited 31/05/2005).

118 The Waterworks Act 1932, South Australia, assented on 30 November 1932. It has been amended many times; the version of Reprint No. 11 available at http://www.austlii.edu.au/au/legis/sa/consol_act/wa1932164/ (last visited 31/05/2005).

119 The Sewerage Act 1929, South Australia, assented on 30 October 1929. It has been amended many times; the version of Reprint No. 6 available at http://www.austlii.edu.au/au/legis/sa/consol_act/sa192996/ (last visited 31/05/2005).

120 The Environmental Protection Act 1997, South Australia, assented on 27 October. It has been amended many times; the version of Reprint No. 16, dated 24 November

management of water resources. On the whole, the present statutory phase is mainly regulated by the Water Resource Act 1997 and/or the Natural Resources Management Act 2004, although this phase shall be managed essentially by the latter act.

5.3.2 Definition of water resource

In order to manage water resources well, the Natural Resources Management Act 2004 gives the meaning and components of water resource a full interpretation in section 3 "Interpretation". According to section 3(1),

> "water resource" means a watercourse or lake, surface water, underground water, stormwater (to the extent that it is not within a preceding item) and effluent;
>
> "watercourse" means a river, creek or other natural watercourse (whether modified or not) in which water is contained or flows whether permanently or from time to time and includes—
>
> (a) a dam or reservoir that collects water flowing in a watercourse;
>
> (b) a lake through which water flows;
>
> (c) a channel (but not a channel declared by regulation to be excluded from the ambit of this definition) into which the water of a watercourse has been diverted;
>
> (d) part of a watercourse;
>
> (e) an estuary through which water flows;
>
> (f) any other natural resource, or class of natural resource, designated as a watercourse for the purposes of this Act by an NRM plan;
>
> "lake" means a natural lake, pond, lagoon, wetland or spring (whether modified or not) and includes—
>
> (a) part of a lake; or
>
> (b) a body of water designated as a lake—
>
> (i) by an NRM plan; or
>
> (ii) by a Development Plan under the Development Act 1993;
>
> "surface water" means—
>
> (a) water flowing over land (except in a watercourse)—
>
> (i) after having fallen as rain or hail or having precipitated in any other manner; or

---

2003, available at http://www.austlii.edu.au/au/legis/sa/consol_act/epa1993284/ (last visited 31/05/2005).

(ii) after rising to the surface naturally from underground;

(b) water of the kind referred to in paragraph (a) that has been collected in a dam or reservoir;

"underground water" means—

(a) water occurring naturally below ground level;

(b) water pumped, diverted or released into a well for storage underground;

effluent means domestic wastewater or industrial wastewater;

domestic wastewater means—

(a) water used in the disposal of human waste; and

(b) water used for personal washing; and

(c) water used for washing clothes or dishes; and

(d) water used in a swimming pool;

"industrial wastewater" means water (not being domestic wastewater) that has been used in the course of carrying on a business (including water used in the watering or irrigation of plants) that has been allowed to run to waste or has been disposed of or has been collected for disposal".[121]

Further, in the Act, "a water resource includes all aspects of a water resource, including the water, organisms and other components and ecosystems that contribute to the physical state and environmental, social and economic value of a water resource".[122]

From these detailed definitions of the terms concerning water resources, one may well understand the meaning of water resources, and this may prove a valuable model for a water-shortage jurisdiction such as China to learn from. However, the Act does not confer ownership of water itself within a natural source upon anyone, but rather sets out the right to access and take water, which will now be discussed in more detail.

5.3.3 Property right on water resources

5.3.3.1 The right to take water

In Australia, South Australia is the first state that introduced water trading on the River Murray as early as 1983,[123] and a relatively ordered trading system

[121] See section 3 (1), Natural Resources Management Act 2004.

[122] See section 3 (2) (b), Natural Resources Management Act 2004.

[123] See The Department of Water, Land and Biodiversity Conservation, *Annual Report 2001-2002*, p. 25.

has been established. However, in South Australia, it is necessary to differentiate owning physical water from having a right to take water from a natural source. "Water itself, while still forming a part of the natural resource, can not be said to be 'owned' by anybody. It is considered a public commodity. However, once water has been legally captured from its natural source and taken under a person's control, that person could then be said to own that particular water."[124]

The right to take water from a natural source in South Australia is now completely regulated by statute. In accordance with section 124 of the Natural Resources Management Act 2004, the author would like to categorize the right to take water into the following six types of right. However it should be noted that there are particular provisions of law to limit these rights.

(1) The right to take water from a non-prescribed source for any purpose by a person who has lawful access to a watercourse, lake or well;[125]
(2) The right to take surface water from land for any purpose by the occupier of the land;[126]
(3) The right to take water from a prescribed water source with a water licence;[127]
(4) The right to take water from a prescribed source or area by the occupier of land from, "(i) a watercourse that adjoins or runs through the land, or (ii) a lake that adjoins or is on the land; or (iii) a well that is on the land";[128]
(5) The right to take surface water from a prescribed area "by the occupier of land from the land, and is used by the occupier for domestic purposes or for watering stock (other than stock subject to intensive farming)";[129]
(6) The right to take water "for the purposes of drinking or cooking by the person who takes it or by a person to whom he or she gives the water if the rate at which the water is taken does not exceed the rate prescribed by regulation".[130]

However, not all of these rights to take water are personal property, but only the right to take water from a prescribed water source with a water licence, in terms of law, "[a] licence (including the water allocation of the licence) is personal property vested in the licence and will pass to another person" in accordance with the law.[131]

---

124 *Supra* note 111, p. 2.
125 See section 124 (1), Natural Resources Management Act 2004.
126 *Id.*, section 124 (2).
127 *Id.*, section 124 (3).
128 *Id.*, section 124 (5) (a).
129 *Id.*, section 124 (5) (b).
130 *Id.*, section 124 (7).
131 *Id.*, section 146 (11).

5.3.3.2 Water licence

" Water licence means a licence granted under Chapter 7 Part 3" of the Natural Resources Management Act 2004 authorising the holder to take (or to hold) water from a watercourse, lake or well or to take (or to hold) surface water from a surface water prescribed area and includes a licence granted under that Part endorsed with a water (holding) allocation".[132] There are two types of water licence:

(1) A licence endorsed with a water (taking) allocation. Water (taking) allocation "means the quantity of water that the licensee is entitled to take and use pursuant to" a corresponding water licence.[133] The holder of this type of water licence is authorised "to take water from the prescribed watercourse, lake or well or to take surface water from the surface water prescribed area specified in the licence".[134]

(2) A licence endorsed with a water (holding) allocation. Water (holding) allocation "means the quantity of water that the licensee is entitled to request that the Minster convert to a water (taking) allocation" in accordance with the law.[135] The holder of this type of water licence is not authorised to take water, but the licence will enable the holder to make a request to the Minister to convert the allocation to a water (taking) allocation in accordance with the law.[136]

The application for a licence The applicant submits an application form to the Minister and pays the fee provided by the relevant regulation.[137]

The refusal of a licence Under the following circumstances, the Minister may refuse to grant a water licence,

> "(a) if it is not possible to endorse a water allocation on the licence consistently with the relevant water allocation plan; or
>
> (b) if, in the opinion of the Minister, the water to be taken pursuant to the licence is so contaminated that its use would create a risk to the health of people or animals; or
>
> (c) to a person, or to the associate of a person, who formerly held a licence that was cancelled under this Act; or
>
> (d) to a person who has acted in contravention of this Act; or
>
> (e) on any other reasonable ground." [138]

---

[132] *Id.*, section 3 (1).
[133] *Id.*
[134] *Id.*, section 146 (1) (a).
[135] *Id.*, section 3 (1).
[136] *Id.*, section 146 (1) (b).
[137] *Id.*, section 146 (2).

The content of a licence A licence must have the following content: (a) a specification of the water resource from which the water is to be taken, (b) a specification of the part or parts of the resource from which the water may be taken when the licence is a licence endorsed with a water (taking) allocation, (c) an endorsement with a water allocation in detail, (d) the conditions of the endorsement, and (e) a specification, whether necessary upon the Minister's decision, which indicates "intervals at which the Minister may vary the conditions of the licence".[139]

The stability of a licence Firstly, a licence is in force until it is terminated by or under the Natural Resources Management Act 2004.[140] That is to say, a licence may be terminated in accordance with the law. Secondly, a licence "is subject to conditions prescribed from time to time by regulation".[141] Thirdly, the Minister in the following situations may vary a licence:

"(a) at any time on the application of, or with the consent of, the licensee; or

(b) if the licence provides for intervals at which the conditions of the licence may be varied—at those intervals if, in the Minister's opinion, the variation is necessary or desirable to more effectively regulate the use of water from the resource—

(i) in accordance with the relevant water allocation plan and this Act; or

(ii) in accordance with the objects of the River Murray Act 2003 or the Objectives for a Healthy River Murray under that Act; or

(c) at any time if there has been an alteration to the water allocation plan for the water resource specified in the licence and the variation is necessary, in the opinion of the Minister, to prevent the licence from being inconsistent (as to the quantity of water allocated or the basis on which it is allocated) or seriously at variance (as to the licence conditions) with the plan; or

(d) on or after 1 July 2006, insofar as the variation is being made on account of the operation of section 146(7) in order to provide for the allocation of water under the licence on a basis that does not relate to the use of water for a crop; or

(e) at any time if the variation is to impose or vary a condition of a licence that relates to a water resource within the Murray-Darling Basin and the Minister is of the opinion that the variation is appropriate or desirable to prevent, reduce or address damage to the River Murray; or

(f) if the Minister is authorised to do so by section 158 or section 162; or

---

[138] *Id.*, section 146 (3).
[139] *Id.*, section 146 (6) (a), (b), (c), (d) and (f).
[140] *Id.*, section 146 (6) (e).
[141] *Id.*, section 146 (6) (6).

(g) at any time in order to reflect a reduction in the water allocation of the licence by the Minister under section 155 or section 156; or

(h) under a scheme established under Division 5." [142]

However, the relevant licensee is entitled to appeal to the ERD Court against the variation of his or her licence under section 147 (1) (b), (c) or (e).[143]

The transfer of a licence In accordance with section 38, subject to the relevant water allocation plan, there are three types of transfer of licence: (a) the transfer of a licence, including its water allocation, from a licensee to another person, (b) the transfer of the whole or part of the water allocation of a licence from a licensee to another licensee or to the Minister, (c) The transfer of a licence or the whole or part of a water allocation to or by South Australian Water, and (d) the transfer of a licence or the whole or part of a water allocation from a licensee to any other person or to the Minister under an Interstate Water Entitlements Transfer Scheme.[144]

The transfer may be absolute as well as for a limited period,[145] but a limited period transfer will slip back automatically to the transferor or the transferor's licence when that period ends.[146] Any of these types of transfer must be officially completed with an approval from, or variations of the transferring and receiving licences by, the Minister. In detail, a transfer under type (a), (c) or (d) must be completed with the approval from the Minister,[147] and a transfer belonging to type (b) have to be achieved by the variation of the Minister.[148]

Concerning a transfer under type (a), the Minister may decline to grant approval for the transfer to a person on the same justification as those on which the Minister would reject to grant an application by that person for a licence.[149] Regarding a transfer under type (b), the Minister may reject to vary the licences concerned if the transfer is to a person who has acted in contravention of the Natural Resources Management Act 2004.[150] As to a transfer under other types, the Minister may refuse to endorse it if the licensee contravenes any condition of the relevant licence.[151]

The procedure for the transfer of a licence According to section 158, the process for the transfer may be described as follows:

---

[142] *Id.*, section 147 (1).
[143] *Id.*, section 147 (3) (b).
[144] *Id.*, section 157 (1).
[145] *Id.*, section 157 (2) and (4).
[146] *Id.*, section 157 (7).
[147] *Id.*, section 157 (2), (6) and (8).
[148] *Id.*, section 157 (4).
[149] *Id.*, section 157 (3).
[150] *Id.*, section 157 (5).
[151] *Id.*, section 157 (6).

(1) The licensee submits an application form, which shall be accompanied by the regulated fee and the licence or licences affected due to the application;[152]

(2) Where the Minister thinks appropriate, he/she may order an evaluation of the effect of granting the application to be made at the expense of the applicant and by a professional assigned or agreed by the Minister;[153]

(3) If an application involves a licence that concerns a water resource within the Murray-Darling Basin and falls within a class prescribed by the particular regulations, before making a decision, the Minister must, (i) consult with the Minister who is in charge of the administration of the River Murray Act 2003, and (ii) comply with the latter's relevant directions, if any.[154]

(4) When granting an application, the Minister may, (i) decrease the water allocation of the transferred licence, (ii) in the case of an application for the transfer of the water allocation of a licence to another licence, require that a quantity of water of the receiving licence is less than that of the transferring licence, (iii) modify any condition of the licence transferred or the receiving licence, and (iv) take any other necessary action in accordance with an Interstate Water Entitlements Transfer Scheme.[155]

The basis for making a decision According to section 160 of the Act, the Minister should make a decision to, either grant or refuse approval, on the following bases: (a) where the water allocation is approved on the transferred or receiving licence, the decision ought to be constant with the relevant water allocation plan, or concerning the conditions attached to the licence, must not be critically inconsistent with the relevant water allocation plan; (b) the decision must be taken in the public interest; (c) whenever the licence involves a water resource concerning the Murray-Darling Basin, only after considering the relevant terms, requirements or any resolution, shall the decision be made; and (d) the decision must be in conformity with any other requirements, if any, prescribed by the regulation. However, where a person is noted on the register of water licences as having an interest in a licence, the Minister must not endorse the transfer without the written assent from that person.[156]

From the above discussions on the right to take water and licences, it is clear that, in South Australia, (a) not all rights to water are property rights, and (b) there is a set of norms that regulate the application for, refusal of, content of, stability of, transfer of, and procedure for the transfer of a licence, as well as the

---

[152] *Id.*, section 158 (1).
[153] *Id.*, section 158 (2).
[154] *Id.*, section 158 (3).
[155] *Id.*, section 158 (4).
[156] *Id.*, section 160.

basis on which the Minister makes a decision. These may prove a valuable lesson for water-shortage jurisdictions including China.

However, as to the property right of the water that has been taken by a holder of the right to take water, the Natural Resources Management Act 2004 does not regulate, although the water is required to be used under the conditions required by the Act. It is logical to conclude that the water would be regulated by property law or the law of property. That is to say, the water that has been taken from a water natural source should be a subject of property law; however, the owner of this water should use the water in accordance with the law, including the Natural Resources Management Act 2004, or he/she/it will be sanctioned. For example, when a licensee contravenes or fails to comply with a condition of his/her licence, he/she/it is guilty of an offence, and the Minister may cancel, suspend or vary the licence in accordance with the Natural Resources Management Act 2004.

5.3.4 The environmental right to water

Although the environmental right to water is not employed as a term in water legal regime in South Australia, there are substantial provisions dealing with this right. As will be revealed later, it is thought that "water-dependent ecosystems have a legitimate right to water" in accordance with the Natural Resources Management Act 2004, and water for the environment is a vital component of water resources management.

In order to understand the present regulations on the environmental right to water, it is necessary to present a brief review of the policy evolution. In the Water Reform Framework 1994, the introduction of a comprehensive system of water allocations is recommended, and the provision of water for the environment, together with the determination of clearly specified water entitlements and water trading arrangements constitute the reform package.

Following the Water Reform Framework 1994, the National Principles for the Provision of Water for Ecosystem was adopted in 1996 by two Ministerial councils, *i.e.*, the Agricultural and Resource Management Council of Australia and New Zealand (ARMCANZ) and the Australia and New Zealand Environmental and Conservation Council (ANZECC).[157] This document provides a "policy direction on how the issue of water for the environment should be dealt with in the context of general water allocation decisions".[158]

---

[157] ARMCANZ and ANZECC, *National Principles for the Provision of Water for Ecosystem*, available at http://www.netspace.net.au/~jnevill/Env_flows_principles96.doc (last visited 31/05/2005).

[158] *Id.*, p. 7.

In the document, the necessary of reviewing the existing water allocation procedures so as to allow use for agricultural, domestic and industrial purposes as well as for environmental purpose, *i.e.*, ensuring sufficient water allocated to meet environmental need, is recognized, and twelve principles are declared. This document introduces a new term, *i.e.*, water-dependent ecosystems. "Water dependent ecosystems are those parts of the environment, the species composition and natural ecological processes of which are determined by the permanent or temporary presence of flowing or standing water. The instream areas of rivers, riparian vegetation, springs, wetlands, floodplains and estuaries are all water dependent ecosystems."[159] As to the meaning of environmental water requirement, the document defines it as "descriptions of the water regimes needed to sustain the ecological values of aquatic ecosystems at a low level of risk. These descriptions are developed through the application of scientific methods and techniques or through the application of local knowledge based on many years of observation".[160]

These twelve principles declared read as follows:

> "Principle 1 River regulation and/or consumptive use should be recognised as potentially impacting on ecological values.
>
> Principle 2 Provision of water for ecosystems should be on the basis of the best scientific information available on the water regimes necessary to sustain the ecological values of water dependent ecosystems.
>
> Principle 3 Environmental water provisions should be legally recognised.
>
> Principle 4 In systems where there are existing users, provision of water for ecosystems should go as far as possible to meet the water regime necessary to sustain the ecological values of aquatic ecosystems whilst recognising the existing rights of other water users.
>
> Principle 5 Where environmental water requirements cannot be met due to existing uses, action (including reallocation) should be taken to meet environmental needs.
>
> Principle 6 Further allocation of water for any use should only be on the basis that natural ecological processes and biodiversity are sustained (ie. ecological values are sustained).
>
> Principle 7 Accountabilities in all aspects of management of environmental water provisions should be transparent and clearly defined.
>
> Principle 8 Environmental water provisions should be responsive to monitoring and improvements in understanding of environmental water requirements.

159 *Id.*, p. 8.
160 *Id.*

Principle 9 All water uses should be managed in a manner which recognises ecological values.

Principle 10 Appropriate demand management and water pricing strategies should be used to assist in sustaining ecological values of water resources.

Principle 11 Strategic and applied research to improve understanding of environmental water requirements is essential.

Principle 12 All relevant environmental, social and economic stakeholders will be involved in water allocation planning and decision-making on environmental water provisions."[161]

Needless to say, these twelve principles are all directly related to the environmental right to water, and deal with different aspects of this right.

Under the Natural Resources Management Act 2004, a legal framework is established. Within the legal framework, many of the COAG reforms, including those related to water for the environment, are addressed in the following three aspects.

The first aspect is that the environmental right to water is recognised. The object of the Natural Resources Management Act 2004 requires that the water-dependent ecosystems and their associated biological diversity (biodiversity) shall be protected. This Act recognises that the water-dependent ecosystems have a legitimate right to water. Section 7 (1) of the Natural Resources Management Act 2004 articulates the sole object of the Act. It reads as follows:

"The objects of this Act include to assist in the achievement of ecologically sustainable development in the State by establishing an integrated scheme to promote the use and management of natural resources in a manner that—

(a) *recognises and protects the intrinsic values of natural resources*; and

(b) seeks *to protect biological diversity* and, insofar as is reasonably practicable, *to support and encourage the restoration or rehabilitation of ecological systems and processes that have been lost or degraded*; and

(c) provides for the protection and management of catchments and the sustainable use of land and water resources and, insofar as is reasonably practicable, seeks to enhance and restore or rehabilitate land and water resources that have been degraded; and

(d) seeks to support sustainable primary and other economic production systems with particular reference to the value of agriculture and mining activities to the economy of the State; and

---

161 *Id.*, pp. 11-15.

(e) provides for *the prevention or control of impacts caused by pest species of animals and plants that may have an adverse effect on the environment*, primary production or the community; and

(f) promotes educational initiatives and provides support mechanisms to increase the capacity of people to be involved in the management of natural resources." [emphasis added][162]

In addition, due to the articulated fundamental elements of ecologically sustainable development, one may find that the provision for water for the environment is undoubtedly a vital component of the object.[163] Further, in accordance with section 7 (3), twelve principles listed should be applied in achieving ecologically sustainable development. All of these principles are directly or indirectly related to the environmental right to water.[164] It is obvious

---

[162] See section 7 (1), Natural Resources Management Act 2004.

[163] Under section 7 (2) of the Natural Resources Management Act 2004, "ecologically sustainable development comprises the use, conservation, development and enhancement of natural resources in a way, and at a rate, that will enable people and communities to provide for their economic, social and physical well-being while— (a) sustaining the potential of natural resources to meet the reasonably foreseeable needs of future generations; and (b) safeguarding the life-supporting capacities of natural resources; and (c) avoiding, remedying or mitigating any adverse effects of activities on natural resources".

[164] In accordance with section 7(3) of the Natural Resources Management Act 2004, these twelve principles read as follows:

"(a) decision-making processes should effectively integrate both long term and short term economic, environmental, social and equity considerations;

(b) if there are threats of serious or irreversible damage to natural resources, lack of full scientific certainty should not be used as a reason for postponing measures to prevent environmental degradation;

(c) decision-making processes should be guided by the need to evaluate carefully the risks of any situation or proposal that may adversely affect the environment and to avoid, wherever practicable, causing any serious or irreversible damage to the environment;

(d) the present generation should ensure that the health, diversity and productivity of the natural environment is maintained or enhanced for the benefit of future generations;

(e) a consideration should be the conservation of biological diversity and ecological integrity;

(f) environmental factors should be taken into account when valuing or assessing assets or services, costs associated with protecting or restoring the natural environment should be allocated or shared equitably and in a manner that encourages the responsible use of natural resources, and people who obtain

that a more holistic and ecologically sustainable approach to water resources management is adopted under the Natural Resources Management Act 2004.

The second aspect is that the Natural Resources Management Act 2004 sets out different ways for water for the environment to be achieved through various water management plans. For example, the State Natural Resources Management Plan (State NRM Plan) must propose principles and policies for achieving the objects of the Act throughout South Australia,[165] in which the environmental right to water is recognised. A water allocation plan made by a regional NRM board must include an assessment of the quantity and quality of water needed by the water-depended ecosystem, and ensure that an equitable balance is achieved between the environmental, social and economic needs for the water.[166]

The third aspect is that the individuals or bodies involved in the administration of the Natural Resources Management Act 2004, or the public, have a right and are under a duty to provide the water for the environment. For example, (a) wherever a water resource, whether prescribed or not, is in inadequate supply or overused, the Minster may prohibit or restrict the taking of water, or direct that water works be modified to allow water to pass over, under

---

benefits from the natural environment, or who adversely affect or consume natural resources, should bear an appropriate share of the costs that flow from their activities;

(g) if the management of natural resources requires the taking of remedial action, the first step should, insofar as is reasonably practicable and appropriate, be to encourage those responsible to take such action before resorting to more formal processes and procedures;

(h) consideration should be given to Aboriginal heritage, and to the interests of the traditional owners of any land or other natural resources;

(i) consideration should be given to other heritage issues, and to the interests of the community in relation to conserving heritage items and places;

(j) the involvement of the public in providing information and contributing to processes that improve decision-making should be encouraged;

(k) the responsibility to achieve ecologically sustainable development should be seen as a shared responsibility between the public sector, the private sector, and the community more generally;

(l) the local government sector is to be recognised as a key participant in natural resource management, especially on account of its close connections to the community and its role in regional and local planning."

165 See section 74 (2), the Natural Resources Management Act 2004.

166 *Id.*, section 76 (4) (a) (b).

or through them in due process; and "the need for water of the ecosystems that depend on water from the water resource concerned must be taken into account".[167] (b) In order to prevent harm and/or further harm to a water-dependent ecosystem, the Minister may reduce licensed water allocations. [168] (c) When a decision is made under Chapter 7 "Management and protection of water resources", the decision-maker, including the Minister or other person or body, must consider the requirements of the ecosystems that depend on the relevant water resource.[169] (d) Among other things, a by-law relating to water resources made by a regional NRM board must be consistent with the objects of the Natural Resources Management Act 2004.[170]

Public participation may be employed as another example. In the National Principles for the Provision of Water for Ecosystem, Principle 12 states that, all relevant "stakeholders will be involved in water allocation planning and decision-making on environmental water provisions". [171] Moreover, this principle is fully reflected in the Natural Resources Management Act 2004. For example, when creating or amending the State NRM Plan, the Natural Resources Management Council (NRM Council) must, (a) "prepare a draft of the proposal", and (b) "by public notice, give notice of the place or places at which copies of the draft are available for inspection (without charge) and purchase and invite interested persons to make written representations on the proposal within a period specified by the NRM Council".[172]

In the State Water Plan 2000, the environmental right to water is fully reflected and well regulated. Under the State Water Plan 2000, of the four core or key values for water use and management, three are relevant to water for the environment, *i.e.*,

> "Availability of water of sufficient quantity and quality for human use and the environment is fundamental to maintaining and improving the quality of life for all South Australians;
>
> South Australia's water is precious and must be managed and used according to the principles of ecologically sustainable development through which water is used so as to maximise its economic, social and environmental returns on sustainable basis;
>
> South Australia's water resources must be managed in an integrated manner with all other natural resources." [173]

---

[167] *Id.*, section 132 (2).
[168] *Id.*, section 156 (1) (b).
[169] See section 170, Natural Resources Management Act 2004.
[170] *Id.*, section 171 (5) (a).
[171] See ARMCANZ and ANZECC, *supra* note 152, Principle 12,
[172] See section 74 (8) (a) and (c), Natural Resources Management Act 2004.
[173] See DWR, *State Water Plan 2000*, Vol. 1, p. 25.

In the State Water Plan 2000 (Vol. 1) – Policies for a Sustainable Future, the environmental right to water is mainly discussed in Section 4.4 which is entitled "Managing the Health of Water-Dependent Ecosystems" and six aspects are discussed, *i.e.*, (i) integrated management of waterbodies and associated water-dependent ecosystems, (ii) water for the environment, (iii) riparian zone management, (iv) wetlands management, (v) floodplain planning and management, and (vi) estuary management; further, the goal(s) for each aspect is/are established and corresponding principles are introduced.[174] Analysing these, it emerges that the environmental right to water is given higher priority, on the one hand. For example, of the general principles for the integrated management of waterbodies and associated water-dependent ecosystems, Principle 2 requires that "[p]revention of degradation to water-dependent ecosystems in good condition should, in general, be considered as a higher priority than the rehabilitation of degraded water-dependent ecosystems"; and Principle 6 further declares that, "[w]here water-dependent ecosystems have become degraded, the goal for management activities should recognise the potential for recovery and generally for rehabilitation as a first priority"; further Principle 7 stipulates that "[e]nvironmental water provisions are not tradable".[175]

Of the principles for providing water for the environment, Principle 2 believes that, "[w]ater allocation and management decisions must take a precautionary approach by first ensuring that natural ecological processes and biodiversity of water-dependent ecosystems are maintained. It follows that further allocation of water for new consumptive uses, and any other new water resource developments, must ensure that ecological values are protected".[176]

Moreover, of the principles for riparian management, Principle 3 provides that, "[p]rotection of refuge areas and maintenance of water connections up and down the watercourse must be given priority due to the highly variable flow patterns of many South Australian watercourses", and Principle 9 requires priorities for rehabilitation should be generally set on certain bases.[177]

On the other hand, the role of public participation is also taken into consideration. Principle 5 of the general principles for the integrated management of waterbodies and associated water-dependent ecosystems provides that, "[a]ll stakeholders involved in development and resource use planning for water-dependent ecosystems need to ensure that (a) the ecological values of water-dependent ecosystems are identified and then protected, and (b) that there are clear objectives for management and rehabilitation projects".[178]

---

174 *Id.*, pp. 36-47.
175 *Id.*, p. 38.
176 *Id.*, p. 40.
177 *Id.*, p. 43.
178 *Id.*, p. 38.

Further, Principle 10 of the principles for providing water for the environment provides that, "[a]ll relevant stakeholders should be involved in planning in relation to environmental water provisions".[179]

In other parts of the State Water Plan 2000, the environmental right to water is also integrated. For example, the "provision for environmental water needs" is listed as one of the four goals for the management and allocation of water resources,[180] and transfer of water rights between catchment is generally not supported because of the potential environmental impacts but with the exception of the transfer being within the ecological limits of the taking and receiving environments.[181]

It is not difficult to conclude that, the environmental right to water is well regulated in South Australia water law. Firstly, the environmental water needs are recognised. Secondly, the environmental water needs are given proper priority. Thirdly, the law confers rights and duties on the relevant individuals and organs to protect the environmental right to water. A water-shortage country, including China should take note of these.

5.4 CONCLUSION

Although water rights could be analysed in three aspects, different backgrounds may lead to different understandings. Consequently, the emphasis and/or extent may be different. As to the term "water resources", water law within both jurisdictions not only defines it clearly, but also further the component of water resource. Additionally, in South Africa, the term "water use" is also regulated. Water resource and water use in South Africa is subject to strict regulation under the water law, more attention is paid to the administration of the government, and it is therefore hard to say that there is a property right on water resources in a sense of market economy. On the contrary, the situation in South Australia appears different, where water right is understood as a right to take water from a natural source, a wide range of individuals may apply for a water licence to take water, and some types of licence are treated as a property right and may be transferred under the conditions imposed by law.

Regarding the human right to water, in South Africa, the apartheid system had meant that most people had no right to access water to meet their basic human needs in law as well as in practice. On the contrary, although, like South Africa, South Australia was under colonial rule, fewer indigenous people and relatively more democratic law and policy have not made the human right to water a vital issue. On the subject of the environmental right to water, South Africa as well as South Australia pays more attention to this, because of severe

[179] *Id.*, p. 40.
[180] *Id.*, p. 47.
[181] *Id.*, p. 51.

aquatic environment issues, and gives priority to the environmental water demand to different extents because of different situations.

Needless to say, the experiences in these two jurisdictions are of great value for a water-shortage jurisdiction, particularly China, in designing a good water rights system or improving an existing water rights system. This will be discussed in the next chapter, Chapter 6 "Comments and Observations on Water Rights System".

6

Comments and Observations on Water Rights System

Following upon the discussions in the previous four chapters, using a comparative and analytical approach, this chapter will firstly conclude the principles and structure recommended to be employed in designing a perfect water rights system, or improving an existing one to ideal, including in China. Secondly, it will examine more closely the key elements of the approach adopted in this work, namely the definition of water resource, the property right on water resources, the human right to water and the environmental right to water, as the important elements of a water rights system, with particular reference to China. Finally, certain specific revisions will be suggested to the China Water Law 2002.

ISBN: 1843391082. Published by IWA Publishing, London, UK.

6.1 PRINCIPLES AND STRUCTURE FOR AN IDEAL WATER RIGHTS SYSTEM

6.1.1 Principles

In order to design an effective and enforceable water rights system, which could promote the reasonable, equitable, efficient and sustainable utilisation of water resources, the author believes the following eight principles should be employed as a foundation.

Water Resource

Principle One. Water resource is not only a social good but also an economic good.

As discussed in Chapter 2, water resource serves multifunctional roles.[1] Water is vital to all living things, as well as to the environment, and is utilised in industries to promote economic development.[2] Therefore, both social and economic values should be recognized under a water rights system. A good water rights system should be designed to regulate water resources and/or water utilisation, and, at the same time, should balance the differing values of water resources. If the social value of water resources alone is taken into consideration, there is a good chance that water resources will be wasted during development and utilisation, and the system will be impossible to implement in our modern society and market economy. However, in a particular jurisdiction, the emphasis on these elements may differ in some way, *e.g.*, the aspects of the human right to water and the environmental right to water in South Africa, or the aspects of the property right relating to some rights to take water and the environmental right to water in South Australia. Each of these jurisdictions provides valuable lessons for establishing a new or improving an existing water rights system.

Principle Two. Water resource falls into the category of things that must be regulated by/in law.

From the discussion in the previous chapters, it is clear that, from the different approaches, or in the different jurisdictions (*e.g.*, China, South Africa and South Australia), the definitions for water resource vary, because different people understand this term in different ways, and/or different jurisdictions have different situations regarding water resources.[3] Thus, there is a range of

---

1 See the relevant discussion in 2.1 of Chapter 2.

2 *Id.*

3 See the relevant discussion in 2.1.4 of Chapter 2, in 3.3.1 of Chapter 3, 5.2.2.1 and 5.3.2 of Chapter 5.

possible approaches to regulate the definition and utilisation of water resource in law.

However, in a specific jurisdiction, the definition of water resource must be specified by law in accordance with the principle of the rule of law. Water resource is regarded as a thing in law within all jurisdictions examined here. As mentioned above, however, in different jurisdictions, water resource in its natural state may be subject to different laws because of the various approaches adopted. In South Africa, for example, the National Government is the public trustee of the nation's water resources.[4] In New Mexico, USA, "[a]ll natural waters flowing in streams and watercourses, whether such be perennial, or torrential, within the limits of the state of New Mexico, belong to the public and are subject to appropriation for beneficial use",[5] while in Colorado, "[t]he water of every natural stream…is hereby declared to be the property of the public, and the same is dedicated to the use of the people of the state, subject to appropriation as hereinafter provided".[6] In China, water resource is owned by the state and is subject to the State administration.[7] In South Australia, although the legal status of water resource in its natural state is not provided for, water resource is subject to the government administration under the Natural Resources Management Act 2004.[8] In France, "[w]ater forms part of the common assets of the nation. … Water use belongs to all, within the framework of the laws and regulations as well as prior established rights".[9]

The author would like to propose that water resource should be subject to different laws, *e.g.*, the law of property, human rights law and environmental law, and as a key method of control, the law of property could be employed as a basis in order to promote the sustainable utilisation of water resources. In China, the application of property law under the civil law system to regulate water resources would be better able to deal with certain important issues, *e.g.*, the confusion relating to the ownership and other property rights concerning water resources, inappropriate regulations on the different kinds of water, and the import and export of fresh water, as discussed in more detail in Chapter 3 above.[10] As an object governed by property law, water resource should be considered an object of ownership as well as of usufruct under condition(s) stipulated by law or the administration. This would be the recommended approach.

---

[4] See section 3 (1), National Water Act 1998.

[5] See art. 1.1, New Mexico Water Law. NMSA 1978, §72-1-1 (1907).

[6] See art. XVI, Section 5, The Constitution of Colorado. Available at http://i2i.org/Publications/ColoradoConstitution/cnart16.htm (last visited 30/05/2004).

[7] See arts. 3 and 12, China Water Law 2002.

[8] See sections 7 and 8, Natural Resources Management Act 2004.

[9] See art. 1, French Water Act 1992.

[10] See the relevant discussion in 3.4 of Chapter 3.

Property Right on Water Resources

Principle Three. Water resource in its natural state is a thing that should be regulated under the "ownership" concept.

In certain jurisdictions, water resource in its natural state under law belongs to all people within the respective jurisdiction, and the State is under a trusteeship to manage it, while not, in fact, owning them. One expert considers that under "an ideal national water law framework", "[w]ater should be owned by the public, administered by the government equitably and for the benefit of that public".[11] However, in our modern society, this does not appear to reflect social reality. Firstly, the State trusteeship over water resources as well as the public ownership on water resources has to be implemented through the administration of the regulated allocation of water resources. Under this approach, the government exercises ownership on behalf of the public. Secondly, the word "public" may be confusing, for the issues about what the term "public" means and what it encompasses (regional, national, international) are difficult to resolve. Thirdly, these two forms are in conflict with the private ownership of water, *de facto* or *de jure*. In reality, water (resources) is sometimes consumed thoroughly or completely. In other words, it may have evaporated or become waste where effective treatment for reuse is too costly. Under these circumstances, how can we talk about trust or public ownership? The author would, therefore, like to suggest that the State trust over water resources could be employed merely in ethics or in spirit, but the form of ownership on water resources is used in legal practice.

The Human Right to Water and the Environmental Right to Water

Principle Four. The human right to water and the environmental right to water shall be recognised explicitly.

Water resource shall not only be the object of property law for its economic value, but shall also be the object of human rights as well as the environmental right for its social value. Human beings have developed due to the evolution of the environment, and economic development benefits human beings. From this viewpoint, the human right to water and the environmental right to water have their natural value. This explains why the human right to water and the environmental right to water are both explicitly or implicitly acknowledged in international law as well as in most domestic jurisdictions, *e.g.*, China, South Africa and South Australia. This should be pursued more effectively under the water rights system in water-shortage jurisdictions including China.

---

[11] See Allan, A. (2003) A Comparison between the Water Law Reforms in South Africa and Scotland: Can a Generic National Water Law Model Be Developed from These Examples? *Natural Resources Journal* **43** (2), 419-489, p.485.

Integration and Administration of Water Rights

Principle Five. The State is under a duty to allocate water resources in a sustainable, equitable and reasonable way, and to ensure that it is utilised equitably, reasonably, effectively and efficiently.

The State ownership over water resources must not be an absolute ownership. The State is under a duty to allocate water resources to meet basic human needs and minimum environmental needs, and to promote social economic development in accordance with the principles of equity, reason and sustainable development. This is reflected in China, South Africa and South Australia. However, in practice, this might be difficult to realise fully and thus the State must develop effective monitoring systems to ensure the successful implementation of these fundamental principles

Principle Six. The property right on water resources should be employed as a tool to protect the human right to water and the environmental right to water, and also as a method to promote the economic development and effective and efficient utilisation of water resources.

Water resource is a type of limited natural resources nowadays. On the one hand, in most cases, it cannot be utilised directly or without being drawn or abstracted. On the other hand, modern society is under the operation of a market system to an increasing extent, and it is necessary to employ the law of property to adjust the utilisation and supply of water resources in certain aspects, in order to ensure that the three aspects of water rights, *i.e.*, the human right to water, the environmental right to water and the property right on water resources, can reach a state of harmony.

Principle Seven. Water rights shall be administered by law.

Throughout the twentieth century, water development in most countries, including the USA, China and the former USSR, relied on government funding and management.[12] In particular, alongside the emergence of modern administrative law since the Second World War,[13] this situation has continued,

---

[12] See Gleick, P. H., et al. (1998) *The World's Water 1998-1999*, Island Press, Washington, D. C., pp. 235-244; Gleick, P. H., et al. (2000) *The World's Water 2000-2001*, Island Press, Washington, D. C., p. 7.

[13] Scott, D. and Felix, A., ed. (1997) *Principles of Administrative Law*, Cavendish Publishing, London, p. 41. It seems that there is a general definition of administrative law accepted in France, USA and UK; See Phillips, O. H. and Jackson, P. (1987) *Constitutional and Administrative Law* (7th ed.), Sweet & Maxwell, London, p. 10. Bradley, A. W. and Ewing K. D. think that, "[a] formal definition of administrative law is that it is a branch of public law concerned with the composition, procedures, powers, duties, rights and liabilities of the various organs of governments the are engaged in the administrating public policies"; see Bradley, A. W. and Ewing K. D. (2003) *Constitutional and Administrative Law*

although controversy over water privatisation has existed since the end of the last century, and some jurisdictions have adopted relevant measures to promote the privatisation of water resources and/or the water services provision. At present, rights to water exist in many areas of social reality, and, as a result, conflict often occurs. Thus, coordination is needed. However, it is difficult for the holders of the different rights to resolve the conflict by themselves, and the State administration of rights is necessary, and should constitute an essential element of a water rights system. No jurisdiction has been found by the author in the contemporary world that has not enacted laws for the government administration of water resources, so as to coordinate different kinds of water needs. Without appropriate and effective administration, the exercise of water rights may prove problematic, and conflicts or disputes may arise which might result in water resources being developed in an unsustainable way. This must be prevented.

Principle Eight. The design of a national water rights system shall be consistent with the relevant rules of international law.

International law is formed with the States' consent in most cases, although subjects other than the States exist. Without international law being obeyed, the world will be without due order, and the States, as well as other subjects, will be in a state of severe conflict. Therefore, every State shall fulfil its international obligations while exercising its rights and sovereignty.

6.1.2 Structure

By following the above eight principles, with an appropriate delimitation of water resource, a scientific system for sustainable integrated water resources management should reasonably balance the multiple functions of water resources, properly take into consideration all competing water needs, and equitably harmonise the economic, social and environmental values of water resources. From a rights-based approach and in a structural sense, an ideal water rights system in a particular jurisdiction should regulate three aspects of water rights, namely, the property rights on water resources, the human right to water and the environmental right to water, based on the foundation of a suitable definition of water resources according to the specific situations in the jurisdiction. These factors will be discussed in the following four sections.

(13th ed.), Longman, Harlow, p. 631. But in China, the scope of administrative law is different from the above concept, because there is a branch of economic law independent from administrative law; see Wouters, P. K., et al. (2004) The New Development of Water Law in China, *Water Law Review* 7 (2) 243-307.

6.2 DEFINITION OF WATER RESOURCE

As pointed out in Chapter 3, Article 2 of the China Water Law 2002 employs the terms "water" and "water resource" simultaneously. Therefore, problems arise, namely, (a) is there any difference between "water" and "water resource"?, (b) if the answer is no, why are two terms employed? and (c) what is the nature of the property right to bottled water bought by a particular individual? In accordance with Article 2(1) of the China Water Law 2002, water resource "includes surface water and groundwater". As to the question, what is "surface water" and "ground water"? neither the law itself nor other legal documents offer corresponding definitions.

In comparison with the definitions under the National Water Law 1998, South Africa, and the Natural Resources Management Act 2004, South Australia, it is obvious that:

Firstly, in the China Water Law 2002, there is not a special article to define or interpret the key terms employed. Such a special article exists in the National Water Act 1998, South Africa, as well as in the Natural Resources Management Act 2004, South Australia.[14]

Secondly, each of these three water laws contains a provision to define water resource in general, *i.e.*, "[w]ater resources referred to in this Law includes surface water and groundwater" in the China Water Law 2002, " 'water resource' includes a watercourse, surface water, estuary, or aquifer" in the National Water Act 1998, South Africa, and "water resource means a watercourse or lake, surface water, underground water, stormwater (to the extent that it is not within a preceding item) and effluent" in the Natural Resources Management Act 2004, South Australia.[15]

Thirdly, the meaning of the components of water resource are not given in the China Water Law 2002, but are mostly provided both in the National Water Act 1998, South Africa and in the Natural Resources Management Act 2004, South Australia.[16]

Fourthly, the general definition of water resource in the China Water Law 2002 is not only inconsistent with other provisions of the Law itself but also with other legal documents.[17] On the contrary, such an issue has not been found in the water law of South Africa and South Australia.

Finally, in the coverage of substantive content, "water resource" as defined under the National Water Act 1998, South Africa, as well as under the Natural

---

[14] See section 1, National Water Act 1998; section 3 (1), Natural Resources Management Act 2004.

[15] See art. 3, China Water Law 2002; See section 1(1) (xxviii), National Water Act 1998; section 3 (1), Natural Resources Management Act 2004.

[16] See the relevant discussion in 3.3.1 of Chapter 3, 5.2.2.1 and 5.3.2 of Chapter 5.

[17] See the relevant discussion in 3.3.1 of Chapter 3.

Resources Management Act 2004, South Australia, is far broader than that under the China Water Law 2002. According to Chinese water law as a whole, one may conclude that "water resource" includes surface water, groundwater, hydraulic power resources and water transport resources. However, excluding hydraulic power resources and water transport resources, "water resource" is more comprehensive in the National Water Act 1998, South Africa, as well as in the Natural Resources Management Act 2004, South Australia. Under the National Water Act 1998, South Africa, "water resource" includes a watercourse, surface water, estuary, or aquifer. Under the Natural Resources Management Act 2004, South Australia, "water resource" includes a watercourse or lake, surface water, underground water, stormwater under stipulated provisions and effluent.

Under international law or domestic law, a jurisdiction is entitled to exercise its power and make its own laws or legislation, including enacting laws or regulations about water resources. Agenda 21 provides repeatedly that all States "according to their capacity and available resources" are to implement their own system of water resources management. However, in the delimitation of water resource, the following aspects should be given special and careful attention: (a) it is helpful to define water resource as well as other important terms in one article or one section in order to avoid confusion on the scope of water resource; (b) water resource not only includes water itself, but also the watercourse or lake, aquifer, estuary etc., which embody water, as well as hydraulic power resources, water transport resources, etc., being resources originating from water, together with its related surroundings. A jurisdiction, however, is entitled and can give a scope to water resource according to its own geographical situation.

As to Chinese water law, the author would like to suggest that the term "water resource" should be defined appropriately. In general, the definition of water resource given in The New Encyclopaedia Britannica is a good reference, *i.e.*, water resource refers to "any of the entire range of natural waters that occur on the Earth, regardless of their state (*i.e.*, vapour, liquid, or solid) and that are of potential use to humans".[18] From the viewpoint of the technique of legislation, the concept of water resource under the Natural Resources Management Act 2004, South Australia, provides a good model for China. On the basis of the scope of water resource in South Australia, combined with the special situation in China, such as being rich in hydraulic power and water transport resources, the author would like to suggest the following definition for water resource:

---

[18] See *The New Encyclopaedia Britannica (Micropaedia)* (15th ed.) (Vol. 12), p. 518.

> Water resource consists of watercourse or lake, surface water, underground water, water under legal control or possession of legal subject, effluent, hydraulic power resources and water transport resources.
>
> Watercourse means a river, creek or other natural watercourse (whether modified or not) and includes (a) a dam or reservoir that collects water flowing in watercourse; and (b) a lake through which water flows; and (c) a channel (but not one declared by regulation to be excluded from the ambit of this definition) into which the water of a watercourse has been diverted; and (d) part of a watercourse.
>
> Lake includes natural lake, pond, lagoon, wetland or spring (whether modified or not) and includes (a) part of a lake; and (b) a body of water declared by regulation to be a lake.
>
> Surface water means (a) water flowing over land (except in a watercourse), (i) after having fallen as rain or hail or having precipitated in any other manner; or (ii) after rising to the surface naturally from underground; and (b) water of the kind referred to in (a) that has been collected in a dam or reservoir.
>
> Underground water includes, (a) water occurring naturally below ground level, and (b) water pumped, diverted or released into a well for storage underground.
>
> Effluent includes, (a) domestic wastewater, *i.e.*, water used in the disposal of human waste, water used for personal washing, water used for washing clothes or dishes, and water used in a swimming pool; or (b) industrial wastewater, *i.e.*, water (not being domestic wastewater) that has been used in the course of carrying on a business (including water used in the watering or irrigation of plants) that has been allowed to run to waste or has been disposed of or has been collected for disposal.

The author believes that these more precise definitions may improve the water rights system in China.

6.3 PROPERTY RIGHT ON WATER RESOURCES

In South Australia, water as property may be categorised as, (a) water forming a part of the natural resources, which is a public commodity, and (b) water under a person's control after being legally captured from its natural source; and some rights to take water are property rights. The author believes, however, that a reasonable property law system, together with a proper administrative legal system applied to water resources, should take into consideration all the possible steps during the process of the utilisation of water resources.[19] The author has realized that services relating to the process of the utilisation of water resources, *e.g.*, water supply services, water treatment services, and wastewater treatment services, etc., may be involved. However, the author considers it is more

[19] As to the process of the utilisation of water resources, see 2.1.2.2 of Chapter 2.

appropriate to differentiate water resources from water services properly for the following three reasons, although sometimes water resources and water services may be combined to a certain extent in some situations, for example, the utilisation of water for export. (a) Water resource is a kind of natural resources, which each State is empowered by sovereignty to administer within its jurisdiction before it becomes a free transferable commodity in the global market. (b) Water resource is a kind of good, and, in general, the law regulating goods is different from the law administering services in any domestic legal system. (c) Under the WTO legal regime, which now includes China,[20] the international trade of goods is subject to General Agreement on Tariffs and Trade, but the international trade of services is under adjustment by the General Agreement on Trade in Services.

As previously mentioned, Chinese legal system primarily belongs to the civil law family,[21] and it is necessary to define water resource as a property in accordance with the *numerus clausus* doctrine of property law, according to which, all property rights should be clearly defined by law. Only after the property right on water resources has been well defined in accordance with the specific circumstances of China, can water resources be allocated and utilised in an equitable, reasonable, efficient and sustainable way. This will make it easier, as well as more convenient, for the public to comprehensively understand and enjoy the right to water and to fulfil their duty concerning water resources, when water resource is clearly provided as an object of property law, when the content of property rights concerning water resources is well delimited.

6.3.1 Ownership on water resources

As stated above, the author favours the State ownership over water resources in its natural state, with private ownership existing in certain conditions.[22] Regarding the method to differentiate water resources in the natural state from those in a non-natural state, it is just a legislative technique. The author would like to recommend the adoption of the following provision:

> Water resources in sources regulated by the State should be owned by the State. Individuals, legal persons or other units may obtain ownership on water resources in accordance with the law.

[20] The PRC became a Member of the WTO on 11 December 2001, See para. 3, *Protocol on Accession of the People's Republic of China - Done at Doha on 10 November 2001 - Notification of Acceptance - Entry into Force*, 20 November 2001, WT/LET/408, available at WTO webpage http://docsonline.wto.org/DDFDocuments/t/WT/LET/408.doc (last visited 20/03/2004).

[21] See the relevant discussion in 3.2.1 of Chapter 3.

[22] See the relevant discussion in 6.1 of this Chapter.

As to the legal content of a property right concerning water resources, the ownership on water resources refers to a comprehensive right in the owner, which includes the right to lawfully possess, utilise, profit from and dispose of his/her/its property. However, individuals, legal persons other than the State, and other units, should only have ownership over physical water, rather than all water resources. As to the methods through which an individual, a legal person other than the State, or any other unit may obtain ownership of water, the author would like to suggest the following, provided that the private ownership of water is subject to conditions imposed by law or the permit or license (hereinafter "licence").

(1) According to the provisions of the law directly

An individual should have ownership over the water to be used to meet his/her basic human needs. However this water is not free to be sold.

Any individual, legal person or other unit may have ownership over rainwater collected on land, or property lawfully owned or used by himself/herself/itself. This water should be free to be sold.

(2) To draw water lawfully under a water resources licence system

Water drawn for consumptive use under a water resources licence system ought to be owned by the licensee. The water would have to be used in accordance with the condition(s) imposed by the licence.

Water drawn to be sold as a commodity in the domestic market under a water resources licence system should be owned by the licensee. This water may be sold freely in the domestic market.

Water drawn to be exported as a commodity under a water resources licence system should be owned by the licensee. This water may be exported.

(3) To import water or to buy water that may be sold

Water imported would be owned by the importer and can be sold by the importer.

Water bought from the importer or licensee, who may sell water freely in the domestic market, ought to be owned by the buyer.

Water drawn by the licensee ought to mean that the water has been drawn and is separated from the water resources owned by the State.

6.3.2 Other types of property rights on water resources

Usufruct

In a situation of the non-consumptive use of water resources, the user ought to get usufruct either under a water resources licence or without such a licence. Usufruct over water resources is the right to enjoy water resources which the user has right to possess, utilise and profit from like the owner himself/herself/itself, but with the responsibility of conserving the substance of it, *i.e.*, under a usufruct, the user may not sell the water.

The right to draw water

From the water law in China as well as South Australia, one may conclude that the right to draw water may be conferred directly by law or under a licence issued by the competent government or its working department.[23] Needless to say, each right to draw water has an economic value. However, it is clear that not all of these rights should be taken as property rights under property law. The example of South Australia shows that only certain rights to draw water could be considered property rights.[24]

In different jurisdictions, the condition(s) for obtaining a licence is/are different.[25] Pertaining to charge(s) for a licence, there are seven main charging scenarios: (a) no charge, (b) a charge for the cost of the licence itself being made, (c) a charge for the cost of the licence being processed and made, (d) a charge for the water resources to be utilised; (e) a charge for the cost of the licence being made and the water resources to be utilised; (f) a charge for the cost of the licence being processed and the water resources to be utilised; and (g) a charge for the cost of the licence being processed and made together with the water resources to be utilised.

As to the licence where there is a charge for the water to be utilised, the licence should be said to have a direct economic or property value. As to a licence under the first three circumstances, no direct economic or property value exists. However, sometimes, only after a licence has been granted would the licensee be prepared to invest in or construct the relevant facilities. In this situation, the economic value of the licence should be recognised, for the relevant facilities would have no proper value in the absence the licence. However, it should be noted that under no circumstances should the licensee have a property right over physical water (resources). In fact, there is only a right to acquire a property right over water (resources), although this kind of right may be regarded as a property right as being of indirect economic value.

Another fact that should be mentioned is that inter-basin water diversions exist and should continue to exist, and some of these diversions are organised or operated by the State directly or indirectly. The relevant regions should therefore pay for the cost of water diversion, which cost may be huge sometimes.

---

[23] See the relevant discussion in 3.4.2 of Chapter 3, and 5.3.3 of Chapter 5.

[24] See the relevant discussion in 5.3.3 of Chapter 5.

[25] See the relevant discussion in 3.4.2 of Chapter 3, and 5.3.3 of Chapter 5.

In order to promote investment in the development, utilisation and conservation of water resources, the right to draw water under a licence should be designed to be transferable or negotiable. However, transfer should be subject to certain conditions regulated by law, and the relevant provisions in the Natural Resources Management Act 2004, South Australia, are of value to be considered. The author would like to suggest that the right to draw water under a licence issued for commercial water use should be transferable, subject to approval from the issuer. In addition, regarding the right to draw water under a licence for non-commercial water use, Article 45(2) of the Water Law of the People's Republic of China (Amendment Draft), which was proposed by the State Council on 26 November 2001, should be accepted, because this provision may promote the efficient utilisation of water resources as well as encourage investment, and, moreover, is helpful in the fulfilment of the economic value of water resources.[26]

6.3.3 Advantage of applying the design in China

Under the design recommended above, which provides for a well-defined and consistent structure for the property rights on water resources, many problems in current Chinese water law, identified in Chapter 3, could be resolved.

The provision that the un-diverted water in its natural state is owned by the State could resolve the ownership issues on waters in reservoirs. Because the State is the owner of the waters in a reservoir owned by a rural collective economic organisation, it is therefore unnecessary to distinguish these waters from the waters of the river in which the reservoir is located. Further, there would be no difference between a reservoir built within the lands owned by a rural collective economic organisation and that built within the lands owned by the State; and no difference between reservoirs built upstream, downstream or midstream, with the exception of the property rights of related land.

As to the issues arsing from the cases of the Yiwu-Dongyang water transaction and the Zhangzhe water transaction, *i.e.*, the holder of the State ownership, the objects of the transactions may be resolved. Under the above design, the waters in the transaction would be owned by the State. Thus, the said buyers would have to obtain water-utilisation licences inititally, and only then, would they be entitled to divert water. Moreover, the object of the sale is the services supplied by the owners or operators of the relevant waterworks.

26 Article 45 (2) of Water Law of the People's Republic of China (Amendment Draft) provided that: "[a]ny unit or individual which has obtained the right to draw water, and has saved water . . . within the duration and quantity of the relevant licence, are entitled to transfer the saved water resources after the approval of the authorities which issued the relevant licence...".

Additionally, the issue of ownership other than the State ownership could also be resolved using the above recommendations. In detail, for the water in ponds or small dams built within contractual areas in the activities of their production under contracted responsibility, the contracting parties other than the State or the governments would have ownership over the water with appropriate restrictions. An example of such a restriction would be that the water could only be utilised in the activities relating to the purposes of the relevant contracts. As to a bottle of water bought by a private individual, logically, this should be owned by that private individual, which is consistent with the recommended approach.

6.4 THE HUMAN RIGHT TO WATER

Under current Chinese water law, especially the China Water Law 2002, the human right to water is not recognised explicitly as a right in law. There are also insufficient provisions to ensure the fulfilment and protection of the human right to water. However, under international law, as one of State Parties of the International Covenant on Economic, Social and Cultural Rights 1966, China is under an obligation to adopt appropriate measures to implement the human right to water.

6.4.1 The human right to water shall be clearly declared in water law

As mentioned earlier, due to the serious water shortages, many thousands of people are suffering from the lack of minimum safe water to meet their basic human needs each year, as a result severe drought, serious pollution and so forth.[27] Although the Chinese government has adopted some policies taken up relevant measures to deal with this problem, the human right to water is, at times, ignored or not paid adequate attention because of its weak status in law. It is therefore necessary to make the human right to water a definite right in law, as is the case in South Africa.

6.4.2 The human right to water shall be given the first priority

According to Article 21(1) of the China Water Law 2002, one may conclude that the human right to water is given the first priority among all water uses. It should be noted, however, that this article is in Chapter Three, which is entitled "Exploitation and Utilisation of Water Resources". Comparing the human rights

[27] See the relevant discussion in 3.5 of Chapter 3.

in South Africa water law, the author would like to propose that the provision which declares the human right to water is given first priority ought to be put in the first chapter, *i.e.*, Chapter One which is entitled "General Provisions". This would mean that the human right to water is given the first priority as a principle, and as one of the key principles, it would be reflected in all other subsequent chapters, *i.e.*, Chapter Two "Planning for Water Resources", Chapter Three "Exploitation and Utilisation of Water Resources", Chapter Four "Protection of Water Resources, Water Areas and Waterworks", Chapter Five "Allocation and Saving Utilisation of Water Resources" and Chapter Six "Settlement of Water Dispute, Enforcement of Law, Supervision and Inspection". Whenever there is no definite provision(s) to deal with matters concerning the human right to water, this principle can be employed to give greater prominence and importance to the human right to water.

6.4.3 The legal content of the human right to water shall be defined

In order to ensure that the human right to water is adequately protected and effectively implemented, it is necessary to provide the legal content of the human right to water. The author recommends that the relevant provisions in General Comment 15 should be transplanted into Chinese water law, *i.e.*,

> The human right to water entitles everyone to sufficient, safe, acceptable, physically accessible and affordable water for personal and domestic uses.[28]

However, some appropriate amendments could be made according to the particular situations in China and in accordance with the technique of legislation. The author, who is knowledgeable of the situation in China, believes that appropriate amendments could be made so as to ensure that an effective human right to water is contained in the Chinese water rights system.

6.5 THE ENVIRONMENTAL RIGHT TO WATER

Under current Chinese water law, especially the China Water Law 2002, the environmental right to water is not acknowledged expressly as a right in law. Further, there is inadequate stipulation to guarantee the implementation and protection of the environmental right to water. However, China is the contracting party of some relevant international treaties, *e.g.*, the Ramsar Convention,[29] the World Heritage Convention,[30] the UN Framework

---

[28] See para. 2, General Comment 15.

[29] On 31 March 1992, the Chinese government deposited the instrument of accession of the Ramsar Convention. Up to date, there are 21 important wetlands in China that

Convention on Climate Change,[31] the UN Convention on Biological Diversity,[32] on the one hand. On the other hand, China has committed to soft international law documents, such as the Rio Declaration on Environment and Development,[33] Agenda 21,[34] the UN Millennium Declaration,[35] and the Johannesburg Declaration on Sustainable Development.[36] Therefore, in

---

have been listed as the Wetlands International Sites; see *The List of Wetlands of International Importance*, available at http://www.ramsar.org/sitelist.doc (last visited 30/05/2004).

30 On 22 November 1985, the Standing Committee of the NPC ratified the World Heritage Convention; see the Standing Committee of the NPC, The Decision on the Ratification of the Convention concerning the Protection of the World Cultural and Natural Heritage, official Chinese version, *Gazette of the Standing Committee of the NPC* (1985) 6, 39. Up to date, there are 30 sites in China that have been listed on World Heritage List; see UNESCO, *the World Heritage List*, available at UNESCO webpage http://whc.unesco.org/pg.cfm?cid=31(last visited 30/05/2004).

31 On 11 June 1992, the Chinese government signed the UN Framework Convention on Climate Change. Shortly after the signature, the Standing Committee of the NPC ratified the Convention on 7 November 1992. See the Standing Committee of the NPC, The Decision on the Ratification of the United Nations Framework Convention on Climate Change, official Chinese version, *Gazette of the Standing Committee of the NPC* (1992) 6.

32 On 11 June 1992, the Chinese government signed the UN Convention on Biological Diversity. Shortly after the signature, the Standing Committee of the NPC ratified the Convention on 7 November 1992. See the Standing Committee of the NPC, The Decision on the Ratification of the United Nations Convention on Biological Diversity, official Chinese version, *Gazette of the Standing Committee of the NPC* (1992) 6.

33 For example, the PRC was represented at the UN Conference on Environment and Development, which was held at Rio de Janeiro, Brazil, from 3 to 14 June 1992. In the conference, the Rio Declaration on Environment and Development was adopted. See UN Doc. A/CONF.151/26 (Vol. IV).

34 For Example, the PRC was represented at the UN Conference on Environment and Development, which was held at Rio de Janeiro, Brazil, from 3 to 14 June 1992. In the conference, the Agenda 21 was adopted. See UN Doc. A/CONF.151/26 (Vol. IV). Further, the Chinese government has enacted its own agenda for the implementation of Agenda 21, and set up a special institution to deal with this issue; see the relevant discussion in 4.1 of Chapter 4.

35 See The UN Country Team in China, *Millennium Development Goals: China's Progress*, Office of the United Nations Resident Coordinator in China, 2004.

36 For example, the PRC was represented at the World Summit on Sustainable Development, which was held in Johannesburg, South Africa, from 26 August to 4 September 2002. In the summit, the Johannesburg Declaration on Sustainable Development was adopted. See UN Doc. A/CONF.199/20*. See also the UN Country Team in China, *id*.

accordance with international law, China should take suitable steps to recognise and enforce the environmental right to water.

6.5.1 The environmental right to water shall be clearly acknowledged in water law

With severe shortages of water resources in certain regions, the environment in China has been suffering serious degradation. It is appreciated that the eco-environmental water demand is acknowledged as one water demand,[37] and the Chinese government has adopted some measures to mitigate the adverse impact on the environment. However, the measures taken are predominantly administrative and *ad hoc*, and there are not enough legal documents to supply a normative and effective mechanism to deal with the environmental right to water. There are cases where the government leaders are unwilling to pay attention to the environmental needs for water resources, and so assorted negative influences occur, which are difficult to rectify.[38] The author therefore believes that the experience in South Africa should be an example, and that it is essential to declare the environmental right to water clearly and explicitly in law.

6.5.2 The environmental right to water shall be given a proper priority

In accordance with Article 21(1) of the China Water Law 2002, in the exploitation of and utilisation of water resources, after the domestic water demands of urban and rural inhabitants have been satisfied, agricultural, industrial and eco-environmental water demands as well as navigational requirements shall be taken into consideration.[39] It is natural to conclude that the priority of the environmental right to water is secondary to the human right to water, agricultural water demand, and industrial water demand. Clearly, the priority given to the environmental right to water is divergent from its importance, and far from adequate. The author believes that the provisions relating to the environmental right to water in South Africa is a good example for China to follow. The author would like to consider that: (i) the environmental right to water in China should be given the first priority together with the human right to water; or (ii) at least, it should be given the second priority immediately after the human right to water; or (iii) as last choice, it

[37] See art. 21 (1), China Water Law 2002.

[38] See State Environmental Protection Administration of the PRC, *China Environmental Status Gazette 2003*, its Chinese version available at http://www.zhb.gov.cn/eic/649368303189360640/index.shtml (last visited 20/04/2004).

[39] See art. 21 (1), China Water Law 2002.

should be given the second priority together with agricultural water demand immediately after the human right to water. It is suggested that this should be provided for in the first chapter of the China Water Law 2002. The environmental right to water should be guaranteed a higher priority in law, and must be taken as one principle, and, as one of the principles, it is necessary to be echoed in all other substantial chapters. It is also suggested that, when there is no definite provision(s) to deal with matters pertaining to the environmental right to water, this principle should be applied, as a matter of priority.

6.5.3 The general legal content of the environmental right to water shall be defined

In order to ensure that the environmental right to water is protected and duly implemented, it is crucial to define its legal content. In this respect, the author would like it to be written as follows, in accordance with the emerging principles of international law:

> (a) The environment has an indispensable right to get appropriate water with minimum quantity and proper quality to maintain its heath. This right shall be protected by law.
>
> (b) Human beings and their communities at various levels are under a duty or responsibility to ensure the realization of the environmental right to water. Appropriate rights or obligations shall be entitled or imposed on the individuals and the communities by law.

6.5.4 Concrete provisions shall be provided

The law and policy relating to the water-dependent ecosystem in South Australia provide a useful precedent for a water-shortage jurisdiction such as China, in order to make concrete provisions on the environmental right to water. The author believes that at least the following two aspects should be carefully considered.

Firstly, a system should be established to define the minimum proper water for some vital or important parts of the environment, such as important rivers, lakes, water regions, groundwater, aquifers or wetlands etc.

Secondly, in order to assure the minimum proper water for some vital or important parts of the environment, whenever any emergency provided by law occurs, the government should adopt efficient measures, including stopping the licensee's right to taken water without compensation.

6.6 RECOMMENDATIONS

In accordance with the above comments, the author would like to propose the following specific amendments to the China Water Law 2002.

6.6.1 On Article 2(2)

To modify Article 2 (2) as follows and treat it as a separate article:

> Water resource consists of watercourse or lake, surface water, underground water, water under the legal control or possession of a legal subject, effluent, hydraulic power resources and water transport resources.
>
> Watercourse means a river, creek or other natural watercourse (whether modified or not) and includes,
>
> (a) a dam or reservoir that collects water flowing in a watercourse; and
>
> (b) a lake through which water flows; and
>
> (c) a channel (but not one declared by regulation to be excluded from the ambit of this definition) into which the water of a watercourse has been diverted; and
>
> (d) part of a watercourse.
>
> Lake includes natural lake, pond, lagoon, wetland or spring (whether modified or not) and includes,
>
> (a) part of a lake; and
>
> (b) a body of water declared by regulation to be a lake.
>
> Surface water means,
>
> (a) water flowing over land (except in a watercourse), (i) after having fallen as rain, snow or hail or having precipitated in any other manner; or (ii) after rising to the surface naturally from underground; and,
>
> (b) water of the kind referred to in (a) that has been collected in a dam or reservoir.
>
> Underground water includes,
>
> (a) water occurring naturally below ground level, and,
>
> (b) water pumped, diverted or released into a well for storage underground.
>
> Effluent includes,
>
> (a) domestic wastewater, *i.e.*, water used in the disposal of human waste, water used for personal washing, water used for washing clothes or dishes, and water used in a swimming pool; or,

(b) industrial wastewater, *i.e.*, water (not being domestic wastewater) that has been used in the course of carrying on a business (including water used in the watering or irrigation of plants) that has been allowed to run to waste or has been disposed of or has been collected for disposal.

6.6.2 On Article 3

Article 3 should be modified as follows:

Water resources in sources regulated by the State belongs to all people of the State. The government is the public trustee of the nation's water resources, and in form the State has ownership on the water resources.

The State ownership on water resources shall be exercised by the State Council on behalf of the State. The State Council must ensure that water resources are developed, utilised, conserved, protected, managed and controlled in a sustainable and equitable manner, for the benefit of all persons and in accordance with its constitutional mandate.

Individuals, legal persons or other units may obtain ownership or other property rights on water resources in accordance with the provisions of this Law and other relevant laws.

6.6.3 Obtaining of ownership on water resources

To add the following content and treat it as a separate article:

Individuals, legal persons and other units may obtain ownership on water by the following methods, however private ownership of water is subject to conditions imposed by law or the water resources licence.

(1) according to the following provisions directly

Any individual may have ownership on the water to be used to meet his/her basic human needs. However this water is not free to be sold.

Any individual, legal person or other unit may have ownership on rainwater collected on land, property lawfully owned or used by himself. This water is free to be sold.

Any individual, legal person or other unit may have ownership on water in accordance with the provisions of other laws.

(2) to draw water lawfully under the water resources licence system

Water drawn for consumptive use under the water resources licence system may be owned by the licensee. The water shall be used in accordance with the condition(s) imposed by the licence.

Water drawn to be sold as a commodity in the State under the water resources licence system may be owned by the licensee. The water may be sold freely in the domestic market.

Water drawn for export as a commodity under the water resources licence system may be owned by the licensee. The water may be exported.

(3) to import water or to buy water that may be sold

Water imported shall be owned by the importer and may be sold by the importer.

Water bought from an importer or a licensee who may sell water freely in the State shall be owned by the buyer.

Drawn water means that water has been drawn and is separated from the water resources owned by the State.

6.6.4 Obtaining usufruct on water resources

To add the following content to Chapter One and treat it as a separate article:

In the circumstances of the non-consumptive use of water resources, the user as a licensee may obtain usufruct on water resources under a water resources licence.

Usufruct on water resources is the right to enjoy water resources of which the licensee has the right to possess, utilise and profit from the water resources, but with the obligation to conserve the substance of the water.

6.6.5 Transfer of the right to draw water under a licence

To add the following content to Chapter One and treat it as a separate article:

The right to draw water under a licence may be transferred in accordance with the law.

Within the duration[40] and quantity of a licence issued for commercial water use, the right to draw water under the licence is transferable subject to the approval from the authorities which issued the relevant licence, the licenser.

Within the duration and quantity of a licence issued for non-commercial water use, when the licensee saves water through efficient use, the licensee may transfer the saved right to draw water under the licence, and this transfer is subject to the approval from the authorities which issued the relevant licence.

---

[40] Either in South Africa water law or under South Australia water law, there are provisions regulating duration of a water use or a licence; see the detailed discussion in 5.2.2, 5.3.3 of Chapter 5. Also, under the present water-drawing licence system in China, there is a limitation on the duration of a licence; see the relevant discussion in 3.4.2.1 of Chapter 3.

> The procedure and conditions for transfer of the right to draw water under licence shall be regulated by the State Council.

6.6.6 The human right to water

To add the following content to Chapter One and treat it as a separate article:

> The human right to water entitles everyone to have access to sufficient, safe, acceptable, physically accessible and affordable water for personal and domestic use. The State shall take all appropriate measures to promote the realisation of the human right to water.

6.6.7 The environmental right to water

To add the following content to Chapter One and treat it as a separate article,

> The environment has an indispensable right to get appropriate water with minimum quantity and proper quality to maintain its health. This right is protected by law.
>
> The State shall adopt appropriate measures to ensure the realization of the environmental right to water. Appropriate rights or obligations shall be entitled or imposed on the people's government at different levels, individuals and units.

6.6.8 On Article 4

To modify Article 4 as follows:

> The exploitation, utilisation, saving or protection of water resources, as well as the prevention or control of water disasters, shall be carried out through comprehensive planning with all relevant factors taken into consideration, tackling problems with both ad hoc and radical measures, emphasising multi-purpose utilisation and the achievement of optimum benefits, giving full play to the multiple functions of water resources.
>
> In coordinating different water-uses, livelihood water-use shall be given the first priority, eco-environmental water-use the second priority, agricultural water-use the third priority, and then production, business water-use and other water-uses the lowest priority.

6.6.9 Minimum proper water for the important parts of the environment

To add the following content and treat it as a separate article:

> The State shall set up a system to ensure the minimum proper water for the vital or important parts of the environment, such as important rivers, lakes, water regions, groundwater aquifers, wetlands, etc.
>
> When the minimum proper water for the vital or important parts of the environment prescribed by the State can not be satisfied, the government shall adopt efficient measures, including the right to stop the licensee's right to take water without compensation.

Theoretically, the water rights under or resulting from these amendments to the China Water Law 2002 proposed above could be well administered and protected. With respect to the situation in China, a legal system composed of seven branches is emerging, and water law belongs to the branch of economic law.[41] Furthermore, within the institutional framework under the China Water Law 2002,[42] it is possible for these water rights to be administered properly. By applying the Administrative Licensing Law of the PRC,[43] a licence for the right to draw water may be issued according to due process. Under the Law for Water Pollution Control of the PRC (Revision), the Environment Protection Law of the PRC, and the Environmental Assessment Law of the PRC, the environmental right to water may be taken into careful consideration, and be effectively protected and properly monitored. According to the Administrative Punishment Law of the PRC,[44] which regulates the set-up and implementation procedure of the administrative sanctions on private individuals, a penalty or penalties could be applied against an illegal action under the water law using the administrative methods. In accordance with the Administrative Reconsideration

---

[41] See the relevant discussion in 3.2.3 of Chapter 3.

[42] This institutional framework concerns many executive authorities, such as the MWR, basin administrative authorities, local people's governments, the competent departments of water administration of local people's governments, and other departments of people's governments. See Wouters, P. K., et al. (2004), *supra* note 13, pp. 289-300.

[43] The Administrative Licensing Law of the PRC, adopted at the 4th Session of the Standing Committee of the Tenth NPC on 27 August 2003, entered into force on 1st July 2004. The Administrative Licensing Law of the PRC, official Chinese version, *Gazette of the Standing Committee of the NPC* (2003) 5, 439-448.

[44] The Administrative Punishment Law of the PRC, adopted by the 4th Session of the Eighth NPC on 17 March 1996, came into force on 1st October 1996. The Administrative Punishment Law of the PRC, official Chinese version, *Gazette of the Standing Committee of the NPC* (1996) 3, 3-14.

Law,[45] when a private individual believes that a particular administrative action performed by an administrative authority has infringed his/her/its lawful right(s), he/she/it may launch a process of administrative reconsideration. Property rights, meanwhile, are regulated by the General Principle of Civil Law of the PRC. Additionally, there are procedure laws that allow private individuals to seek remedies, such as the Administrative Procedure Law of the PRC,[46] which regulates the legal actions initiated by private individuals against particular administrative actions, and the Civil Procedure Law of the PRC,[47] which normalises the litigations between equal legal subjects of civil law. Therefore, there exists a comprehensive legal mechanism in place to enable the corresponding State organs or authorities to enforce the laws that regulate water resources, and also administer and protect water rights.

In brief, in this chapter, the principles that should be employed in designing or improving a water rights system are suggested. Based on these principles, the key elements of a water rights system in China are discussed in detail with an international, comparative and analytical approach. Finally, nine specific amendments to the China Water Law 2002 are recommended. In the next chapter, a brief conclusion will be presented with a short commentary to explain the basis for the recommendations set forth.

[45] The Administrative Reconsideration Law of the PRC, adopted at the 9th Session of the Standing Committee of the Ninth NPC on 20 April 1999, entered into force on 1st October 1999. The Administrative Reconsideration Law of the PRC, official Chinese version, *Gazette of the Standing Committee of the NPC* (1999) 3, 225-234.

[46] The Administrative Procedure Law of the PRC, adopted by the 2nd Session of the Seventh NPC on 4th April 1989, came into force on 1st October 1989. The Administrative Procedure Law of the PRC, official Chinese version, *Gazette of the Standing Committee of the NPC* (1989) 2.

[47] The Civil Procedure Law of the PRC, adopted by the 4th Session of the Seventh NPC on 9th April 1991, came into force on the same day. The Civil Procedure Law of the PRC, official Chinese version, *Gazette of the Standing Committee of the NPC* (1991) 3.

7

Conclusions

Based on the previous five chapters, especially Chapter 6, the author would like to provide the following conclusions.

(a) According to hydrology, any water is part of the water cycle. On the one hand, however, only limited physical and reachable freshwater can be of particularly vital or important value to human beings and to certain environments. Sometimes, physical freshwater, together with certain surroundings, can engender other kinds of water resources. On the other hand, human beings may be unable to control or utilise all freshwater as well as these other kinds of water resources for many different reasons. With the development of society, there is a clear trend towards many different water needs, giving rise to more conflict, which situation has increased the complexity and difficulty of management concerning water. The fact is that human beings have to balance many different needs, and solid legal and institutional frameworks are required to address these problems. Further, a proper water rights system must lie at the heart of the framework.

(b) The first matter to be resolved is to determine what should be the object of management, *i.e.*, the scope or definition of water resource. There are two

ISBN: 1843391082. Published by IWA Publishing, London, UK.

factors requiring attention here: (i) The ethic is, in some ways, different from the reality, although the former could be employed to guide the latter. For example, in a policy sense, however, South Africa treats all water in the water cycle as its water resource, but, in law, there is a definite definition of water resource. (ii) The situations in different countries or jurisdictions are generally quite different; therefore, the scope of water resource *de facto* may be different.

(c) The second issue to be dealt with is the need to balance the many and different water demands so as to reasonably realise all the functions of water resources. The property right aspect of water resources may be employed to promote the realisation of all these functions. Not all water needs are essentially of property values, however, those being primarily of others value may be wholly fulfilled with the help of the property right approach to water resources management, as they are basic human needs to water and essential environmental requirements to water. Therefore, an equitable, reasonable and sustainable water rights system should encompass three aspects, *i.e.*, the property rights on water resources, the human right to water and the environmental right to water.

(d) The reality of State sovereignty brings a unique set of rights and duties. Each States has a responsibility to administer water resources within its jurisdiction for the well-being of its people, without damaging other States and/or the environment in accordance with international law.

(e) In designing a domestic water rights system, certain fundamental principles should be followed, although in any specific jurisdiction the sub-fundamental principles may be the same or differ to some extent, because of particular situations. The author recommends that eight basic principles should be pursued in designing a water rights system, *i.e.*,

(1) Water resource is not only a social good but also an economic good;
(2) Water resource belongs to things that must be regulated by/in law;
(3) Water resource in its natural state is a thing that should be regulated under the "ownership" concept;
(4) The human right to water and the environmental right to water shall be recognised explicitly;
(5) The State is under a duty to allocate water resources in a sustainable, equitable and reasonable way, and to ensure that it is utilised equitably, reasonably, effectively and efficiently;
(6) The property right of water should be employed as a tool to protect the human right to water and the environmental right to water as well as a method of promoting the economic development and effective and efficient utilisation of water resources;
(7) Water rights shall be administered by law;

(8) The design of a national water rights system shall be consistent with the relevant rules of international law.

Under these principles, the legal form, substantive content, as well as the priority of water rights, can be well designed so as to construct an appropriate water rights system that effectively addresses the existing issues that impair sustainable development, including in China.

(f) In order to deal with the water rights issues in China, current Chinese water law has to be improved. In a practical sense, after the term "water resource" has been properly defined, the three aspects of water rights, *i.e.*, the property right on water resources, the human right to water and the environmental right to water, may be carefully addressed. In particular, these rights can be clearly recognised, their subjects or holders can be appropriately regulated, and these rights can be well administered. As to the holders of water rights, the State as well as the private individual can be the subjects of the property right on water resources, the individual natural person can be the holder of the human right to water, and the environment can be the holder of the environmental right to water with the present human beings and groups formed by them at various levels being conferred rights and imposed duties to realise the environmental right to water. Regarding the administration of water rights, the executive authorities, in particular, the competent departments of water administration of the people's governments at and above the county level (especially the MWR), play a vital role within the present institutional framework on integrated water resources management,[1] and the judicial organs will protect the water rights as the final remedy under civil procedure law and administrative law.

[1] See Wouters, P. K., et al. (2004) The New Development of Water Law in China, *Water Law Review* 7(2), 243-307, pp.289-300.

ANNEX: Water Law of the People's Republic of China (2002 Revision)

(English translation by Desheng Hu)

(Adopted at the 29th Session of the Standing Committee of the Ninth National People's Congress on 29 August 2002)

CHAPTER ONE GENERAL PROVISIONS

Article 1 This Law is enacted for the purpose of rationally exploiting, utilising, saving or protecting water resources, preventing or controlling water disasters, realising the sustainable utilisation of water resources, and meeting the needs of national economic and social development.

Article 2 This Law shall be applied to the activities of exploiting, utilising, saving, protecting or managing water resources, as well as preventing or

ISBN: 1843391082. Published by IWA Publishing, London, UK.

controlling water disasters, within the territory of the People's Republic of China.

The water resource referred to in this Law includes surface water and groundwater.

Article 3 Water resources shall be owned by the State. The ownership of water resources shall be exercised by the State Council on behalf of the State. The water of a pond owned by or of a reservoir built and managed by a rural collective economic organisation shall be utilised by the organisation.

Article 4 The exploitation, utilisation, saving or protection of water resources, as well as the prevention or control of water disasters, shall be carried out through comprehensive planning with all relevant factors taken into consideration, tackling problems with both ad hoc and radical measures, emphasising multi-purpose utilisation and the achievement of maximum benefits, giving full play to the multiple functions of water resources, and coordinating water-uses between livelihood, production and business, and the eco-environment.

Article 5 The people's government at or above the county level shall strengthen the construction of water conservancy infrastructures, and incorporate this into its planning for national economic and social development at the corresponding level.

Article 6 The State shall encourage units or individuals to exploit and/or utilise water resources in accordance with the law, and protect their legitimate rights and/or interests. Any unit or individual engaged in exploiting or utilising water resources shall be under an obligation to protect water resources in accordance with the law.

Article 7 The State shall implement a water-drawing licence system and a paid use system on water resources in accordance with the law, with the exception of the water of ponds or reservoirs belonging to rural collective economic organisations utilised by these organisations or their members. The competent department of water administration under the State Council shall be in charge of the organisation and implementation of these two systems in the country.

Article 8 In order to establish a water-saving society, the State shall strictly enforce water-saving activities, greatly promote water-saving measures, spread the application of new water-saving technology and/or techniques, and develop water-saving industries, agriculture and/or service industries.

The people's government at various levels shall adopt measures to strengthen the management of water-saving activities, establish a development and promotion system for water-saving technologies, and foster and develop water-saving industries.

Every unit or individual is under an obligation to undertake water-saving activities.

Article 9 For the purpose of protecting water resources, the State shall adopt effective measures to preserve natural flora, to plant trees and/or grow grass, to conserve water sources, to prevent and/or control soil erosion and water-body pollution, and to improve the eco-environment.

Article 10 The State shall encourage and support conducting the research into, spreading and applying advanced science and technology relating to the exploitation, utilisation, saving, protection and management of water resources, as well as that relating to the prevention and control of water disasters.

Article 11 The people's government shall reward any unit or individual who has demonstrated an outstanding achievement in the exploitation, utilisation, saving, protection or management of water resources, as well as in the prevention or control of water disasters.

Article 12 The State shall carry out a water resources administration system which combines basin administration and administrative region administration.

The competent department of water administration under the State Council is in charge of the unified administration and supervision of water resources in the country.

Within their corresponding jurisdictions, the basin administrative authorities set up by the competent department of water administration under the State Council for the major rivers or lakes so designated by the State, shall perform the relevant duties relating to the administration and supervision of water resources stipulated by law or regulations or authorized by the competent department of water administration under the State Council.

Within their corresponding administrative regions, the competent department of water administration of a local people's government at or above the county level is in charge of the unified administration and supervision of water resources within their jurisdictions.

Article 13 Other relevant departments under the State Council are in charge of the corresponding undertakings of exploitation, utilisation, saving or protection of water resources in conformity with the respective responsibility assigned to them.

Other relevant departments of a local people's government at or above the county level are in charge of the corresponding undertakings of exploitation, utilisation, saving or protection of water resources in conformity with the respective responsibility assigned to them within their administrative regions.

CHAPTER TWO PLANNING FOR WATER RESOURCES

Article 14 The State shall formulate a national water resources strategy plan.

For exploiting, utilising, saving or protecting water resources as well as preventing or controlling water disasters, unified plans should be formulated with the river basin or region as the basic unit. The plans shall be classified into basin plans and region plans. The former include basin comprehensive plans and basin special plans, and the latter include region comprehensive plans and region special plans.

The comprehensive plan mentioned in the preceding paragraph refers to a general plan for the exploitation, utilisation, saving and protection of water resources, as well as the prevention and control of water disasters, in the light of economic and social development needs, as well as the present situation of the development and utilisation of water resources. The special plan mentioned in the preceding paragraph refers to a plan for any special subject including flood control, water-logging prevention, irrigation, navigation, water supply, hydropower generation, bamboo or log rafting, fishery, the protection of water resources, water and soil conservation, the prevention and control of sedimentation and desertification, water-saving activity, etc.

Article 15 Within a basin, any region plan shall be in conformity with the basin plan concerned, and any special plan shall be in conformity with the comprehensive plan concerned.

Any basin comprehensive plan, the region comprehensive plan, or the special plan closely concerning the utilisation of land shall be coordinated with the relevant planning for national economic and social development, the general plans for land use, the general urban plans and the environment protection plans, and shall take into consideration the needs of various regions and different industries.

Article 16 When a plan is formulated, a comprehensive scientific investigation, survey and assessment of water resources must be undertaken. Such an investigation, survey and assessment shall be organised by the competent department of water administration of the people's government at or above the county level, in conjunction with other relevant departments at the corresponding level.

The people's government at or above the county level shall pay special attention to the establishment of information systems for hydrology and water resources. The competent departments of water administration of the people's government at or above the county level and the basin administrative authorities shall strengthen the dynamic monitoring of water resources.

The basic hydrological data shall be accessible to the public in accordance with the relevant provisions of the State.

Article 17 The basin comprehensive plan for a basin of the major rivers or lakes so designated by the State shall be formulated by the competent department of water administration under the State Council, in conjunction with other relevant departments under the State Council and the people's government(s) of the relevant province(s), autonomous region(s) or municipality(ies) directly under the Central Government, and shall be submitted to the State Council for approval. Regarding the basin comprehensive plan and the region comprehensive plan for a river or lake which runs across different provinces, autonomous regions or municipalities directly under the Central Government other than major rivers or basins shall be formulated by the relevant basin administrative authority(ies), in conjunction with the competent departments of water administration and other departments of the people's governments of the relevant province(s), autonomous region(s) and municipality(ies) directly under the Central Government. After being examined and commented on by every people's government of the relevant province(s), autonomous regions and municipalities directly under the Central Government, these plans shall be submitted to the competent department of water administration under the State Council for examination and verification. The competent department of water administration under the State Council shall invite comments from the relevant departments under the State Council, and then submits these plans to the State Council or its authorized department for approval.

The basin comprehensive plan and region comprehensive plan for the basin of a river or lake other than that mentioned in the preceding paragraph shall be formulated by the competent department of water administration of the local people's government concerned at or above the county level, in conjunction with the relevant departments and relevant local people's governments. The competent department of water administration shall submit such a plan to the people's government at the corresponding level or its authorized department for approval, and to the competent department of water administration at the next higher level for record.

The special plans shall be formulated by the relevant departments of the people's government at or above the county level. After inviting comments from other departments at the corresponding level, the relevant departments shall submit such plans to the people's government at the corresponding level for approval. The plans for flood control and those for water and soil conservation shall be formulated and approved in accordance with the relevant stipulations of the Flood Control Law and the Law on Water and Soil Conservation respectively.

Article 18 A plan must be implemented strictly once it has been approved.

Whenever an approved plan needs to be amended, the amendment must be approved by the authority that originally approved the plan in accordance with the formulation procedure applicable to the plan.

Article 19 Any waterwork must be constructed in conformity with the relevant basin comprehensive plan. When a water work is constructed in one of the major rivers or lakes so designated by the State, or in one of the rivers or lakes which run across different provinces, autonomous regions or municipalities directly under the Central Government, before its feasibility study is submitted for approval, the relevant basin administrative authority shall examine whether or not the construction is in conformity with the basin comprehensive plan and give its opinion. Regarding the construction of a waterwork in other rivers or lakes, it is for the competent department of water administration of the relevant local people's government at or above county level to examine and comment on. Where a waterwork involves flood control, its construction shall be carried out in accordance with the relevant provisions of the Flood Control Law. Where a waterwork involves other regions or industries, the construction unit must invite comments in advance from the regions and departments concerned.

CHAPTER THREE EXPLOITATION AND UTILISATION OF WATER RESOURCES

Article 20 The exploitation and/or utilisation of water resources shall follow the principle of promoting benefits while eliminating disasters, take into consideration the interests of upstream and downstream, of the left and right banks and of all regions concerned, fully realize the comprehensive benefits of water resources, and conform to the overall arrangements for flood control.

Article 21 In the exploitation and/or utilisation of water resources, the water-use demands for the livelihood of urban and rural inhabitants shall be satisfied first, while the agricultural, industrial and eco-environment water-use demands, as well as navigation demands etc., shall also be taken into consideration.

In the exploitation and/or utilisation of water resources in arid and/or semi-arid areas, eco-environment water-use demands shall be fully taken into consideration.

Article 22 When an inter-basin water diversion project is to be carried out, a comprehensive plan and scientific study must be undertaken, and an all-round consideration is to be given to water-use demands in the basin from which the water is to be diverted as well as in the basin into which the water is to be diverted, while damage to the eco-environment shall be prevented.

Article 23 When reasonably administering the exploitation and comprehensive utilisation of water resources within their respective jurisdictions, the local people's governments at various levels shall consider their respective actual local conditions of water resources, follow the principles of the unified deployment and exploitation of surface water and groundwater, combining tapping new sources and saving the existing water resources, giving priority to saving the existing water resources, and processing and recycling effluent.

The formulation of plans for national economic and social development and of urban comprehensive plans, as well as the overall arrangement of important and large construction projects, shall be in conformity with the local conditions of water resources and the requirements of flood control, and scientific study shall be undertaken. In areas deficient in water resources, urban development and the development of high water-consumption industries, agriculture and/or services shall be restricted.

Article 24 In areas short in water resources, the State shall encourage the collection, exploitation and/or utilisation of rainwater and brackish water, as well as the utilisation and desalination of seawater.

Article 25 The local people's government at various levels shall strengthen its leadership over the work of irrigation, waterlogging drainage and water and soil conservation, and promote the development of agricultural production. In areas prone to salinization-alkalization and waterlogging, measures shall be undertaken to control and/or lower the groundwater table.

When rural collective economic organisations and/or their members, in accordance with the law, invest in the construction of waterworks on the corresponding land owned by the rural collective economic organisations and/or on contracted land, the waterworks and/or the water stored in these waterworks shall be managed and reasonably utilised according to the principle of "the investor manages and benefits from the waterwork".

The relevant rural collective economic organisation(s), before constructing a reservoir, shall obtain approval from the competent department of water administration of the relevant local people's government at or above the county level.

Article 26 The State shall encourage the exploitation and/or utilisation of hydraulic power resource. On rivers rich in hydraulic power, multi-purposes and cascade development shall be undertaken in a planned manner.

In the construction of hydropower stations, the eco-environment shall be protected, while the requirements for flood control, water supply, irrigation, navigation, bamboo and log rafting, fishery, etc. shall be taken into consideration.

Article 27 The State shall encourage the exploitation and/or utilisation of water transport resource. When permanent dams or sluice-gates are built across the migration routes of aquatic organisms, or across rivers navigable and/or suitable for bamboo and log rafting, the construction unit must simultaneously build facilities for the passage of fish and ships or for bamboo and log rafting, or undertake other remedial measures upon approval from the department(s) authorized by the State Council. The construction unit must also make appropriate arrangements for the protection of aquatic organisms, for navigation, and for bamboo and log rafting during the period of construction and the initial water storing, and also bear all of the expenses incurred thereby.

Wherever a non-navigable river or man-made waterway becomes navigable after a dam or sluice-gate is built across it, the construction unit shall simultaneously build facilities for the passage of ships or reserve sites for such facilities to be built.

Article 28 No unit or individual may, while diverting, intercepting, storing or discharging water, infringe the public interests or the legitimate rights and interests of others.

Article 29 The State shall carry out a development-oriented policy with regard to the construction of waterworks that involves the relocation of inhabitants. Following the principle of combining compensation and/or subsidy given at the early stage with assistance and support given at the later stage, the livelihood and production of the relocated inhabitants shall be arranged properly and the legitimate rights and interests of the relocated inhabitants shall be protected.

The relocation of inhabitants shall be conducted simultaneously with the construction of the relevant waterwork. The construction unit shall, taking into consideration the environmental capacity of the relocation area, following the principle of sustainable development, and adopting measures suited to the local conditions, formulate a plan for the relocation of inhabitants. After being approved in accordance with the law, this plan shall be implemented by arrangement with the local people's government concerned. The funds required for the relocation of inhabitants shall be included in the investment plan for the waterwork.

CHAPTER FOUR PROTECTION OF WATER RESOURCES, WATER AREAS AND WATERWORKS

Article 30 When formulating plans for the exploitation, utilisation, or management of water resources, the competent department of water administration of the people's government at or above the county level, the basin administrative authority or other relevant departments shall pay attention to the maintenance of a rational river flow and a rational water level for lakes, reservoirs and/or groundwater, and maintain the natural purification capacity of the water system.

Article 31 Any unit or individual engaged in activities concerning water resources, including the exploitation, utilisation, saving, and protection of water resources as well as the prevention or control of water disasters, etc., shall follow the relevant approved plans. Anyone who breaches the plan and causes a decrease of the use functions of the water areas of rivers or lakes, the over-exploitation of groundwater, the sinking of land surface or the pollution of water bodies, shall be under a responsibility to eliminate the damage.

Any mining unit or construction unit, when its activity of exploiting minerals or constructing any underground project causes the lowering of the groundwater table, the drying up of any water source or the subsidence of ground, shall undertake remedial measures. Where loss or damage is caused to the lives and production of others, such a unit shall make compensation in accordance with the law.

Article 32 In conjunction with the competent department of environment protection administration and other relevant departments under the State Council as well as the people's governments of relevant provinces, autonomous regions or municipalities directly under the Central Government, the competent department of water administration under the State Council, in accordance with the basin comprehensive plans, the protection plan of water resources and the economic and social development needs, shall draft the water function zones for the major rivers or lakes so designated by the State, and submit that draft to the State Council for approval. Regarding the water function zones for other rivers or lakes which run across different provinces, autonomous regions or municipalities directly under the Central Government, these shall be drafted by the relevant basin administrative authorities in conjunction with the competent departments of water administration, competent departments of environment protection administration and other departments of the relevant people's governments at the provincial level. After these drafts have been examined and commented on by every relevant people's government at the provincial level, the competent department of water administration under the State Council, in conjunction with the competent department of environment protection administration under the State Council, shall examine them, and then submit them to the State Council or its authorized department for approval.

Regarding the water function zones for rivers or lakes other than those stipulated in the preceding paragraph, the competent departments of water administration of the local people's governments at or above the county level, in conjunction with the competent department of environment protection administration and other relevant departments at the corresponding level, shall draft them, and then submit them to the people's government at the corresponding level or its authorized department for approval, and to the competent department of water administration and the competent department of environment protection administration at the next higher level for record.

The competent department of water administration of the people's government at or above the county level and/or the basin administrative authority, in accordance with the water quality requirement for and the natural purification ability of the water body of a water function zone, shall check and define the waste-accepting ability of the water area of the zone, and make a proposal about the limitation on the total amount of discharge of waste into the water area to the competent department of environment protection administration.

The competent department of water administration of the local people's government at or above the county level and/or the basin administrative authority shall monitor the water quality of the water function zone. When they find that the total amount of discharge of the specified key waste exceeds the control limitation, or that the water quality of a water function zone does not meet the water quality requirement for the water use function of that water area, they shall report promptly to the relevant people's government for the latter to undertake control measures, and inform the competent department of environment protection administration concerned.

Article 33 The State shall establish a drinking-water-source protection zone system. The people's governments of provinces, autonomous regions and municipalities directly under the Central Government shall delimit drinking-water-source protection zones, undertake measures to prevent the drying up of water sources and the pollution of water bodies, and ensure the safety of drinking water for both urban and rural inhabitants.

Article 34 The construction of a sewage outlet in a drinking-water-source protection zone is forbidden.

The construction of a new sewage outlet, or the reconstruction or expansion of an existing one, along rivers or lakes, shall be subject to the permission of the competent department of water administration or the basin administrative authority that has jurisdiction over the matter, and to the approval of the environmental assessment report of the project by the competent department of environment protection administration.

Article 35 When the construction of a project involves including water sources for irrigation and/or irrigation and drainage facilities, or has an adverse effect on the existing water for irrigation or sources of water supply, the construction unit(s) shall undertake the necessary remedial measures. When loss is caused, compensation shall be offered in accordance with the law.

Article 36 In areas where groundwater is over-exploited, the local people's government at or above the county level shall undertake measures to control strictly the exploitation of groundwater. In an area where groundwater is seriously over-exploited, forbidden groundwater-exploiting zones or restricted groundwater-exploiting zones may be delimited upon the approval from the people's government of the relevant province, autonomous region or municipality directly under the Central Government. Where the groundwater in coastal areas is exploited, a scientific study shall be conducted, and the relevant measures shall be undertaken to prevent the sinking of the land surface and the encroachment of seawater.

Article 37 No one may abandon or pile up items, or plant trees or grow long-stalk crops, which may block the passage of flood water, in any river, lake, reservoir, canal or channel.

In a river channel management area, no one may construct any building or structure that may block the passage of flood water, undertake any activity that may adversely affect the stability of the river condition, endanger the safety of the river embankment or block the passage of flood water through river channel.

Article 38 In a river channel management area, the construction of any bridge, wharf or other building or structure which blocks, crosses or borders a river channel, as well as the laying of any pipeline or cable which crosses a river, must be in conformity with the flood control standard and other relevant technical requirements prescribed by the State. A project construction plan shall be submitted to the competent department of water administration concerned for examination and approval in accordance with the relevant provisions of the Flood Control Law.

Where an existing waterwork needs to be expanded, modified, dismantled or damaged due to the construction of the project mentioned in the preceding paragraph, the construction unit is responsible for any expense and compensation for loss caused, unless the existing waterwork was constructed illegally.

Article 39 The State shall implement a river channel sand-mining licence system. The measures on implementation of the channel sand-mining licence system shall be stipulated by the State Council.

In a river channel management area, when sand-mining activity may adversely affect the stability of the river conditions or endanger the safety of the river embankment, the competent department of water administration of the relevant people's government at or above the county level shall delimit non-sand-mining areas or prescribe non-sand-mining periods, and publish a notice to this effect.

Article 40 Reclaiming part or whole of a lake for farmland is prohibited. Where such reclamation has been conducted, the lake shall be restored in a planned way in accordance with the flood control standard prescribed by the State.

Reclaiming part or whole of a river channel for farmland is prohibited. Where reclamation is necessary, a scientific study must be conducted, and a reclamation plan shall get consent from the competent department of water administration of the people's government of the relevant province, autonomous region or municipality directly under the Central Government, or from the competent department of water administration under the State Council, before it is submitted to the people's government at the corresponding level for approval.

Article 41 Any unit or individual is under an obligation to protect waterworks, and must not encroach on or destroy river embankments, dykes, flood control facilities, hydrologic monitoring facilities, hydrogeologic monitoring facilities and other relevant facilities.

Article 42 The local people's government at or above the county level shall undertake measures to safeguard waterworks especially dams and river embankments within its administrative region, and to eliminate any dangerous situation within the period it determines. The competent department of water administration shall strengthen the supervision and management of the safety of the waterworks.

Article 43 The state shall protect waterworks. The waterwork management and protection area for any state-owned waterwork shall be delimited in accordance with the stipulations issued by the State Council.

The competent department of water administration under the State Council or basin administrative authority shall delimit the management and protection zones for waterworks administered by it, in consultation with the people's governments of the relevant provinces, autonomous regions or municipalities directly under the Central Government.

Regarding the waterwork management areas and protection duty for waterworks other than those mentioned in the preceding paragraph, these shall be delimited in accordance with the stipulations issued by the people's government of the relevant province, autonomous region or municipality directly under the Central Government.

Within the waterwork protection area, no one is allowed to conduct activities, including blasting, well digging, rock quarrying, soil collecting, etc., which may endanger the safety of the waterwork or adversely affect its operation.

CHAPTER FIVE ALLOCATION AND SAVING UTILISATION OF WATER RESOURCES

Article 44 The competent department of development and planning administration and the competent department of water administration under the State Council are in charge of the macro allocation of national water resources. The competent department of water administration under the State Council, in conjunction with other relevant departments, formulates national and trans-provincial medium-term and long-term water supply-demand plans, and these plans shall be executed after being examined and approved by the competent department of development and planning administration under the State Council. The competent department of water administration of the local people's government at or above the county level, in conjunction with other relevant departments at the corresponding level, formulates the local medium-term and long-term water supply-demand plans, in accordance with the long-term and medium-term water demand-supply plan at the next higher level and the actual regional situation. These plans shall be executed after being examined and approved by the competent department of development and planning administration at the corresponding level.

A medium-term or long-term water supply-demand plans shall be formulated on the basis of the current water supply-demand situation, corresponding national social and economic development planning, basin plan(s) and region plan(s), while following the principles of coordinated water supply and demand, comprehensive balance, ecosystem protection, strict saving and rational tapping of water sources.

Article 45 When regulating and storing run-off, as well as allocating water quantity, a water quantity allocation scheme shall be formulated with the basin as the unit and in accordance with the relevant basin plan and the medium-term and long-term water supply-demand plan.

A water quantity allocation scheme, as well as a water regulation pre-scheme for drought emergencies, which involves different provinces, autonomous regions or municipalities directly under the Central Government, shall be formulated by the relevant basin administrative authority in consultation with the people's governments of the relevant provinces, autonomous regions or municipalities directly under the Central Government, and these schemes shall be executed after being submitted to and approved by the State Council or its authorized department. Other trans-administrative-region water quantity allocation schemes and water regulation pre-schemes for drought emergencies shall be formulated by the competent department of water administration of the common people's government at the next higher level in consultation with the relevant local people's governments, and these schemes shall be executed after being submitted to and approved by the people's government at the corresponding level.

After a water quantity allocation scheme or a water regulation pre-scheme for drought emergencies is approved, the relevant local people's governments must execute it.

A project for the development or utilisation of water resources which is to be constructed across a boundary river concerning different administrative regions, must be in conformity with the approved water quantity allocation scheme of the basin, and the relevant people's governments at or above the county level shall submit the project to the competent department of water administration of their common people's government at the next higher level or the relevant basin administrative authority for approval.

Article 46 The local people's government at or above the county level or the basin administrative authority, in accordance with the approved water quantity allocation scheme and the predicted annual quantity of incoming water, shall formulate an annual water quantity allocation scheme and regulation plan and implement unified water quantity regulation. The local people's government concerned must comply with the above scheme or plan.

The annual water quantity allocation schemes for the major rivers or lakes so designated by the State shall be incorporated into the State annual national economic and social development plan.

Article 47 The state shall implement a water-use system which combines the control of total quantity and quota management.

The competent departments of the people's government of a province, autonomous region or municipality directly under the Central Government shall formulate the water-use quota for each industry within its administrative region, and submit the water-use quota to the competent department of water administration and the competent department of quality supervision and inspection administration for examination and assent, and then the above people's government shall promulgate the water-use quota. The water-use quota shall be submitted to both the competent department of water administration and the competent department of quality supervision and inspection administration under the State Council for record.

The competent department of development and planning administration of the local people's government at or above the county level, in conjunction with the competent department of water administration at the corresponding level, on the bases of water-use quota, economic and technological conditions and the water quantity allocated to its administrative region under the relevant water quantity allocation scheme, shall formulate an annual water-use plan and control the annual total water-use quantity within its administrative region.

Article 48 Any unit or individual who draws water directly from rivers, lakes or ground aquifers shall, in accordance with the State provisions relating to the water-drawing licence system and the paid use system on water resources, apply to the relevant competent department of water administration or basin administrative authority for a water-drawing licence, pay the water resources fee regulated, and then obtain a water-drawing right. However, the drawing of water for a household or for livestock, a small quantity for poultry drinking or other small quantity water drawings, are exempted.

Measures for the implementation of the water-drawing licence system and for the collection and management of water resources fees shall be provided by the State Council.

Article 49 Water-use shall be measured in terms of quantity, and water shall be used in accordance with the approved water-use plan.

For water-use, a water resources fee charging system shall be implemented, under which the water resources fee is charged according to the measurement of quantity, and progressively higher levels of water resources fee are applied for the above-quota water used.

Article 50 The people's governments at various levels shall promote water-saving irrigation methods and water-saving technology, and undertake necessary measures to prevent water leakage in agricultural waterworks for water storage and transport, and therefore increase the efficiency of agricultural water-use.

Article 51 Advanced technology, techniques and equipment shall be adopted in industrial water-use to increase the recycling of the used water, and to raise the ratio of the reuse of water.

The State shall gradually eliminate techniques, equipment and products that are outdated or of high water-consumption. The list for these shall be formulated and published by the competent department of comprehensive economic administration under the State Council in conjunction with the competent department of water administration and other departments concerned under the State Council. The manufacturers, sellers or users shall stop manufacturing, selling and using the techniques, equipment and products listed in the above list within the prescribed period.

Article 52 The urban people's government, taking the local situation into consideration, shall undertake effective measures to promote water-saving household utensils, lower the leakage rate of urban water supply pipelines or networks, increase the efficiency of household water-use, while pay attention to the concentrated treatment of urban effluent and encourage the use of recycled water, and therefore increase the utilisation ratio of recycled effluent.

Article 53 When constructing a new project, or extending or reconstructing an existing project, a scheme for water-saving measures shall be drawn up, and the corresponding water-saving facilities shall be constructed. The water-saving facilities must be designed, constructed and put into operation with the principal part of the project at the same time.

Water supply enterprises or units that build their own water supply facilities shall strengthen the maintenance and management of water supply facilities so as to decrease the leakage and loss of water.

Article 54 The people's governments at various levels shall actively undertake measures to improve the drinking water conditions for both urban and rural inhabitants.

Article 55 When water is supplied by waterworks, the user shall pay a water fee to the water-supply unit in accordance with the provisions of the State. The price for supplied water shall be determined according to the principles of compensation for cost, reasonable profit, better quality justifying higher prices, and an equitable share of the cost. The specific regulations shall be formulated by the competent department of price administration of the people's governments at or above the provincial level, in conjunction with the competent department of water administration and other departments of water-supply administration at the corresponding level in accordance with their functions and powers.

Chapter Six Settlement of Water Dispute, Enforcement of Law, Supervision and Inspection

Article 56 Any water dispute that arises between different administrative regions shall be resolved through negotiation. Where a substantial agreement can not be concluded through negotiation, the water dispute shall be adjudicated by their common people's government at the next higher level. The adjudication shall be obeyed and executed by all parties concerned. Pending the settlement of a water dispute, no party may, in the absence of an agreement concluded by the parties concerned or an approval from their common people's government at the next higher level, construct any project for draining, blocking, drawing, diverting or storing water within a reasonable area on either side of the boundary between the administrative regions concerned, or unilaterally alter the existing water regime.

Article 57 Any water dispute between units or individuals shall be resolved through negotiation. Where a party is unwilling to have the dispute resolved through negotiation, or a substantial agreement can not be concluded through negotiation, the parties concerned may request the local people's government at or above the county level or the department authorized by such a government to mediate the water dispute, or may directly initiate a civil legal action in the people's court. Where the local people's government at or above the county level or the department authorized by such a government can not mediate the water dispute successfully, the parties concerned may initiate a civil legal action in the people's court.

Pending the settlement of a water dispute, no party may unilaterally alter the existing water regime.

Article 58 When handling any water dispute, the people's government at or above the county level, or the department authorized by such a government, has the power to undertake interim disposition measures, to which all sides or parties concerned must comply.

Article 59 The competent department of water administration of the people's government at or above the county level and/or the basin administrative authority shall strengthen the supervision and inspection over the activities against this Law, and investigate and deal with these activities in accordance with the law.

The staff engaging in the supervision and inspection of water administration shall be devoted to their duties and enforce the law impartially.

Article 60 When performing their duties of supervision and inspection stipulated by this Law, the competent department of water administration, the basin administrative authority or their staff engaging in the supervision and inspection of water administration, have the power to undertake the following measures:

(a) To request anyone under inspection to provide the relevant documents, certificates, licences or materials;

(b) To request anyone under inspection to make a statement on issues concerning their compliance with this Law;

(c) To access the production site of anyone under inspection, and carry out an investigation;

(d) To order anyone under inspection to suspend any activity against this Law and to perform his legal obligation.

Article 61 Any unit or individual concerned shall cooperate with the staff engaging in the supervision and inspection of water administration, and must not forbid or impede the staff from performing their duties in accordance with the law.

Article 62 When performing their supervision and inspection duties, the staff engaging in the supervision and inspection of water administration shall produce their enforcement credentials to anyone for inspection.

Article 63 When the people's government at or above the county level, or the competent department of water administration at a higher level, find an activity against the law or a breach of duty being committed in supervision and inspection by the competent department of water administration at the corresponding level or at a lower level, the former shall order the latter to rectify these within a specified period.

CHAPTER SEVEN LEGAL LIABILITY

Article 64 When the competent department of water administration, other relevant department and waterwork management unit as well as their staff, due to accepting property or other benefits from others through taking advantage of their position to, or committing a dereliction of their duty, issue licences or sign endorsement opinions to any unit or individual that does not meet the requirements prescribed by law, fail to allocate water quantity according to the relevant water quantity allocation scheme, to charge a water resources fee according to the relevant regulation of the State, to perform their supervising functions or to investigate and deal with activities that are discovered to be against the law, and any of these activities leads to serious consequences and thus constitutes a crime, any person directly in charge of and any other person directly responsible for the activity shall be investigated for criminal responsibility in accordance with the relevant provisions of the Criminal Law. For any case that is insufficient serious to constitute a crime, any person directly in charge of and any other person directly responsible for the activity shall be given disciplinary sanctions in accordance with the law.

Article 65 Where anyone, in a river channel management area, constructs a building or structure that may block the passage of flood water, undertakes any activity that may adversely affect the stability of the river condition, endanger the safety of a river embankment or block the passage of flood water through the river channel, the competent department of water administration of the people's government at or above the county level, or the basin administrative authority, in accordance with their functions and powers, shall order the illegal actor to cease the illegal activity, dismantle the illegal building or structure, and restore the original condition within a prescribed period. When the illegal actor fails to dismantle or restore, a compulsory dismantling shall be conducted at the expense of the illegal actor, and also a fine of between 10,000 and 100,000 CNY shall be imposed.

Where anyone, without the approval from the competent department of water administration or the basin administrative authority, constructs a waterwork or bridge, wharf or any other building or structure which blocks, crosses or borders a river channel, or lays a pipeline or cable which crosses a river, but the above activity is not regulated under the Flood Control Law, the competent department of water administration of the people's government at or above the county level, or the basin management institution, in accordance with their functions and powers, shall the order illegal actor to cease the illegal activity and apply for the necessary approval or licence within a prescribed period. Where the illegal actor fails to apply or his application is refused, the illegal actor shall be ordered to dismantle the illegal building or structure within a prescribed period. When the illegal actor fails to dismantle, a compulsory dismantling shall be conducted at his expense, and also a fine of between 10,000 and 100,000 CNY shall be imposed.

Where anyone constructs any of the facilities mentioned in the preceding paragraph with the approval from the competent department of water administration or the basin administrative authority, but fails to fulfil the prescribed requirements, the competent department of water administration of the people's government at or above the county level, or the basin administrative authority, in accordance with their functions and powers, shall order the illegal actor to rectify within a prescribed period, and impose a fine of between 10,000 and 100,000 CNY depending on the severity of the illegal activity.

Article 66 Regarding any of the following illegal activities which are not regulated under the Flood Control Law, the competent department of water administration of the people's government at or above the county level, or the basin administrative authority, in accordance with their functions and powers, shall order the illegal actor to cease his illegal activity, remove the obstruction or undertake other remedies within a prescribed period, and also impose a fine of between 10,000 and 50,000 CNY:

(a) Abandoning or piling up items, or planting trees or growing long-stalk crops, which may block the passage of flood water, in a river, lake, reservoir, canal or channel;

(b) Reclaiming part or whole of a lake for cultivated land, or reclaiming part or whole of a river channel for cultivated land without approval.

Article 67 Where anyone constructs a sewage outlet in a drinking-water-source protection zone, the local people's government at or above the county level shall order the illegal actor to dismantle the sewage outlet and restore the original condition within a prescribed period. When the illegal actor fails to dismantle or restore, a compulsory dismantlement or restoration shall be conducted, and also a fine of between 50,000 and 100,000 CNY shall be imposed.

Where anyone constructs a new sewage outlet, or reconstructs or expands an existing one along rivers or lakes without the approval from the competent department of water administration or the basin administrative authority, the competent department of water administration of the people's government at or above the county level, or the basin administrative authority, in accordance with their functions and powers, shall order the illegal actor to cease his illegal activity and restore the original condition within a prescribed period, and impose a fine of between 50,000 and 10,000 CNY.

Article 68 Where anyone manufactures, sells or uses in his production or operation the techniques, equipment and products that are clearly eliminated by the State due to being outdated or of high water-consumption, the competent department of comprehensive economic administration at or above the county level shall order the illegal actor cease the activity of manufacture, sale or use, and impose a fine of between 20,000 and 10,000 CNY.

Article 69 Regarding any of the following illegal activities, the competent department of water administration of the people's government at or above the county level, or the basin administrative authority, in accordance with their functions and powers, shall order the illegal actor to cease his illegal activity and/or undertake remedies within a prescribed period, and impose a fine of between 20,000 and 50,000 CNY. Where the illegal activity involves serious situations, the authority shall revoke any relevant water-drawing licence that has been issued.

(a) Drawing water without approval;

(b) Failing to draw water in accordance with the requirements specified in the water-drawing licence.

Article 70 Where anyone refuses to pay, delays payment of, or is in arrears with the water resources fee, the competent department of water administration of the people's government at or above the county level, or the basin administrative authority, in accordance with their functions and powers, shall order him to make the payment within a prescribed period. When he fails to comply with this order, from the very day of the delay, he shall be charged at a daily rate of 0.2 percent of the total amount, and also be subject to a fine of between one and five times the amount of the water resource fees payable or payable retroactively.

Article 71 Where the water-saving facility of a construction project is put into operation without an approval when the facility is not completed or is completed but fails to meet the State standard, the competent department of water administration of the people's government at or above the county level, or the basin administrative authority, in accordance with their functions and powers, shall order the illegal actor to cease the use of the facility, rectify the activities within a prescribed period, and impose a fine of between 50,000 and 100,000 CNY.

Article 72 Regarding any of the following illegal activities, when it constitutes a crime, investigation shall be conducted for criminal responsibility in accordance with the relevant provisions of the Criminal Law. Where an action is insufficient serious to constitute a crime and not regulated under the Flood Control Law, the competent department of water administration of the people's government at or above the county level, or the basin administrative authority, in accordance with their functions and powers, shall order the illegal actor to cease the illegal activity and undertake remedies, and impose a fine of between 10,000 and 50,000 CNY. Where an illegal activity breaches the Regulations on Security Management and Punishment, the public security organ shall mete out security management punishment to the illegal actor in accordance with the law. When the activity brings losses to others, the illegal actor shall be responsible for compensation in accordance with the law.

(a) Encroaching on or destroying a waterwork, river embankment, dyke or other relevant facility, or destroying a facility relating to flood control, hydrologic monitoring or hydrogeologic monitoring;

(b) Conducting an activity that may endanger the safety of a waterwork or adversely affect the operation of a waterwork within the waterwork protection area, including blasting, well digging, rock quarrying, soil collecting, etc.

Article 73 Where anyone encroaches on, steals or plunders materials for flood control, facilities or equipment for flood and water-logging control, farmland water conservancy, hydrological monitor and survey etc., takes bribes or appropriates State funds or property for disaster relief and rescue, flood control, relocation of and compensation to relocated people, and/or construction of other waterworks, and his activity constitutes a crime, investigation shall be conducted for criminal responsibility in accordance with the relevant provisions of the Criminal Law.

Article 74 Where anyone, in the course of a water dispute or its handling, stirs up unrest, engages in gang fighting, plunders or destroys public or private property, illegally restricts the personal freedom of others, and this activity constitutes a crime, investigation shall be conducted for criminal responsibility in accordance with the relevant provisions of the Criminal Law. When this is insufficient serious to constitute a crime, the public security organ shall mete out security management punishment to the illegal actor in accordance with the law.

Article 75 Regarding one of the following activities in a water dispute between different administrative regions, any person directly in charge of and any other person directly responsible for the activity shall be given disciplinary sanctions in accordance with the law:

(a) Refusing to execute the water quantity water allocation scheme or the water regulation pre-scheme;

(b) Refusing to comply with the unified water regulation;

(c) Refusing to implement the ruling of the people's government at the next higher level;

(d) Pending the settlement of a water dispute, in the absence of agreement concluded by the parties concerned or approval from their common people's government at the next higher level, unilaterally altering the existing water regime against the provisions of this Law.

Article 76 If anyone infringes the public interest or others' legitimate rights and interests due to his diversion, interception (storage) or discharge of water, he shall bear civil responsibility in accordance with the law.

Article 77 The administrative sanctions applied to activities against the river channel sand-mining licence system provided in article 39 of this Law shall be set up by the State Council.

CHAPTER EIGHT MISCELLANEOUS PROVISIONS

Article 78 Where any international treaty or agreement relating to international or border rivers or lakes, concluded or acceded to by the People's Republic of China, contains provisions differing from those in the law of the People's Republic of China, the provisions of the international treaty or agreement shall apply, unless the provisions are ones on which the People's Republic of China has made reservations.

Article 79 The waterwork referred to in this Law includes any kind of works in the rivers, lakes or groundwater sources employed for the exploitation, utilisation, control, allocation and/or protection of water resources.

Article 80 The activities for exploitation, utilisation, protection and management of seawater shall be carried out in accordance with the provisions of other relevant laws.

Article 81 The activities for flood control shall be carried out in accordance with the provisions of the Flood Control Law.

The activities for prevention and control of water pollution shall be carried out in accordance with the provisions of the Law for Water Pollution Control.

Article 82 This Law shall come into force on 1st October 2002.

Select Bibliography

Allan, A. (2003) A Comparison between the Water Law Reforms in South Africa and Scotland: Can a Generic National Water Law Model Be Developed from These Examples? *Natural Resources Journal* **43** (2), 419-489.

Alston, P. (2002) *The United Nations and Human Rights: A Critical Appraisal* (2nd ed.), Clarendon Press, Oxford.

Anderson, M. R. (1998) Human Rights Approaches to Environmental Protection: An Overview, in *Human Rights Approaches to Environmental Protection* (ed. Boyle, A. E. and Anderson, M. R.), Clarendon Press, Oxford, 1-23.

Atiyah, P. S. (1987) *Pragmatism and Theory in English Law*, Stevens & Sons, London.

Barnett, A. D. (1981) *China's Economy in Global Perspective*, The Brookings Institution, Washington D.C.

Bates, R. L. & Jackson, J. A. ed. (1987) *Glossary of Geology*, American Geological Institute, Alexandria.

Beck, R. E., ed. (1991) *Water and Water Rights*, Vol. I., Charlottesville.

Bell, S. & McGillivray, D. (2000) *Environmental law* (5th ed.), Blackstone Press, London.

Birnie, P. W. & Boyle, A. E. (2002) *International Law and the Environment* (2ed ed.), Oxford University Press, Oxford.

Blackstone, W. *Commentaries on the Law of England (1765-1769)*, the University of Chicago Press, 1979.

Borkowski, A. (1997) *Textbook on Roman Law* (2nd ed.), Oxford University Press, Oxford.

ISBN: 1843391082. Published by IWA Publishing, London, UK.

Bradley, A. W. and Ewing K. D. (2003) *Constitutional and Administrative Law* (13th ed.), Longman, Harlow.

Cech, T. V. (2003) *Principles of water resources*, John Wiley & Sons Inc., New York.

Chen, Y. (2002) Civil Law Development: China and Taiwan, *Stanford Journal of East Asia Affairs* (2002) 2, 8-14.

Crabb, J. H. (1995) *The French Civil Code - Revised Edition (as amended to 1 July 1994)*, Fred B. Rothman & Co., Littleton.

D'ENTRÈVES, A. P., *AQUINAS SELECTED POLITICAL WRITINGS*, translated by DAWSON, J. G., Basil Blackwell, Oxford, 1965.

DWAF, *National Water Resources Strategy (First Edition)* (September 2004).

DWAF, *White Paper on a National Water Policy for South Africa* (April 1997).

DWAF, *White Paper on Water Supply and Sanitation Policy* (November 1994).

DWR, *State Water Plan 2000*, Explanatory Documents.

DWR, *State Water Plan 2000*, Vol. 1.

DWR, *State Water Plan 2000*, Vol. 2.

Fan, Z. et al. (1998) Study on the Ecological Environment Evolution and Sustainable Development in Tarim River Basin, in *Proceedings of Seminar on the Water Resources, Environment and Management of Tarim River Watershed (1998)*, 171-177.

FAO (2003), *Review of World Water Resources by Country*, Rome.

Ganer, B. A., ed. (1999) *Black's Law Dictionary* (7th ed.), West Publishing, St. Paul.

Getches, D. H. (1997) *Water Law in a Nutshell* (3rd ed.), West Publishing Co., St Paul.

Getzler, J. (2004) *A History of Water Rights at Common Law*, Oxford University Press, Oxford.

Gillespie, A. (1997) *International Environmental Law, Policy and Ethics*, Clarendon Press, Oxford.

Gleick, P. H., et al. (1998) *The World's Water 1998-1999*, Island Press, Washington, D. C.

Gleick, P. H., et al. (2000) *The World's Water 2000-2001*, Island Press, Washington, D. C.

Gleick, P. H., et al. (2002) *The World's Water 2002-2003*, Island Press, Washington, D. C.

Herschy, R. W. & Fairbridge, R. W., ed. (1998) *Encyclopedia of Hydrology and Water Resources*, Kluwer Academic Publishers, London.

Hohefeld, W. (1913) Some Fundamental Legal Conceptions as Applied in Judicial Reasoning, *Yale Law Journal* **23**(16), 28-32.

Howarth, W. (1992) *Wisdom's Law of Water Courses* (5th ed.), Shaw & Sons Limited, Crayford.

Hu, D. (2004) Research on the Delimitation of Property Rights on Water Resources, *Journal of Henan Finance & Taxation College* **18** (1), 34-37.

Hu, M & Luo, C. (1989) The Ministry of Water Resource and its Power, *The Historical Extreme Floods in China* (1989) 1, 171-187.

Huang, P. C. (2001) *Code, Custom and Legal Practice in China*, Stanford University Press, Stanford.

Jennings, R. & Watts, A., ed. (1992) *Oppenheim's International Law* (9th edition), Longman, Harlow.

Jiang, W. (2000) Research on Water Resources Safety of China in the 21st Century, *China Water Resources* (2000) 8, 41-43.

John Scanlon, et. al. (2004) *Water as a Human Right?* World Conservation Union.

Kuiper, E. (1965) *Water Resources Development: Planning, Engineering and Economics*, Butterworths, London.

Liang, H. (1993) *Civil Law Theories: Cases and Legislations*, (China) Law Press, Beijing.

Linsley, R.K. et al. (1992), *Water-Resources Engineering* (4th ed.) McGraw-Hill, London.

Liyuanhe Water Conservation Management Office (2003) Making Water Rights Clear, Allocating Water to Each Water User, in *Collection of Materials on National Test Regions for Developing Water-saving Society*, China WaterPower Press, Beijing, 2004, 49-55.

Ma, J. (2002) Zhanghe upstream bureau managed the organization of inter-provincial water transaction again, *China Water Resources News*, 19 March 2002.

Marino, M. A. & Simonovic, S. P., ed. (2001) *Integrated Water Resources Management*, Interantional Association of Hydrological Science, Wallingford.

Martin, E. A. (2002) *A Dictionary of Law*, Oxford Press, Oxford.

Matthews, J. A. ed. (2001) *The Encyclopaedic Dictionary of Environmental Change*, Arnold, London.

Mazor, E. (1991) *Applied Chemical and Isotopic Groundwater Hydrology*, Open University Press, Milton Keynes.

McCaffrey, S. C. (2001) *The Law of International Watercourses: Non-Navigational Uses*, Oxford University Press, Oxford.

Montgomery, C. W. (1997) *Environmental Geology* (5th ed.), The McGraw-Hill Companies, Inc., Boston.

Montgomery, C. W. (2003) *Environmental Geology* (6th ed), The McGraw-Hill Companies, Inc., London.

National Bureau of Statistics of the PRC (2003), *China Statistical Yearbook (2003) (Chinese-English version),* China Statistics Press, Beijing.

Niu, F. and Hu, F.(2003) Positively exploring new routes to resolve the water disputes in Zhanhe Basin, *China Water Resources News*, 22 May 2003.

OECD (1989) *Water Resources Management: Integrated Policies*, Paris.

Pearce, R. A. & Stevens, J. (1998) *The Law of Trusts and Equitable Obligations* (2nd ed), Butterworths, London.

Perkins, D. and Yusuf, S. (1984) *Rural Development in China*, The Johns Hopkins University Press, Baltimore.

Pfafflin, J.R., & Ziegler, E. N., ed. (1998) *Encyclopedia of Environmental Science and Engineering* (4th ed.), Gordor and Breach Science Publishers, Australia.

Phillips, O. H. and Jackson, P. (1987) *Constitutional and Administrative Law* (7th ed.), Sweet & Maxwell, London.

Potter, P. B. (2001) *The Chinese Legal System*, Routledge, London.

Qiao, X. (2002) Report of the Law Commission of the NPC on the Revision of the Water Law of the People's Republic of China (Amendment Draft) (26 April 2002), Chinese version, *Gazette of the Standing Committee of the NPC* (2002) 5, 375–378.

Raz, J. (1984) Legal Rights, *Oxford Journal of Legal Study* 4.

Reid, C. T., ed. (1997) *Environmental Law in Scotland* (2ed edition), W. Green, Edinburgh.

Salman, M. A. S. and Mclnerney-Lankford, S. (2004) *The Human Right to Water: Legal and Policy Dimensions*, World Bank Publication, Washington, D.C.

Scott, A. and Coustalin, G. (1995) The Evolution of Water Rights, *Natural Resources Journal* **35** (4), 821-979.

Scott, D. and Felix, A., ed. (1997) *Principles of Administrative Law*, Cavendish Publishing, London.

Deason, J. P., et al. (2001) Water policy in the United States: a perspective, *Water Policy* (2001) 3, 175-192.

Macrory, R. (1985) *Water Law: Principles and Practice*, Longman Professional London.

Shen, H. and Wang, L. (2001) The First transaction on water-use rights was concluded in China, *People's Daily*, 16 February 2001.

Smart, C. (1989) *Feminism and the Power of Law*, Routledge, London.

Solanes, M. & Gonzalea-Villarreal, F. (1999) *The Dublin Principles for Water as Reflected in a Comparative Assessment of Institutional and Legal Arrangements for Integrated Water Resources Management*, Global Water Partnership, Stockholm.

Soussan, J. (1988) *Primary Resources and Energy in the Third World*, Routledge, London.

The UN Country Team in China, *Millennium Development Goals: China's Progress*, Office of the United Nations Resident Coordinator in China, 2004.

Tushnet, C. (1984) An Essay on Rights, *Texas Law Review* 63, 1363.

UNESCO et al. (2003) *The United Nations World Water Development Report: Water for People, Water for Life (Executive Summary)*, UNESCO Publishing/Berghahn Books, Paris.

UNESCO, et al. (2003) *The United Nations World Water Development Report: Water for People, Water for Life*, UNESCO Publishing/Berghahn Books, Paris.

UNESCO/WMO (1992) *The International Glossary of Hydrology* (2nd ed.), Paris.

Vasan, R. S., ed. (1980) *The Canadian Law Dictionary*, Law and Business Publications (Canada) Inc., Don Mills.

Waldron, J., ed. (1984) *Introduction to Theories of Rights*, Oxford Press, Oxford.

Walker, D. M. (1980) *The Oxford Companion To Law*, Clarendon Press, Oxford.

Wang, L. (2001) The first transaction on water rights was concluded in Zhejiang Province, *China Water Resources News*, 10 February 2001.

Wang, L., ed. (2004) *Chinese Civil Code (Draft): Suggestion and Explanation*, China Legal System Press, Beijing.

Wang, S. (2001) A Reply to Wu Guoping, *China Water Resources News*, 28 February 2001.

Wang, S. (2001) Explanation to Water Law of the People's Republic of China (Amendment Draft) (24 December 2001), *Gazette of the Standing Committee of the NPC* (2002) 5, 371–374.

Wang, S. (2003) Working Key-points in Developing Water-saving Society, in *Collection of Materials on National Test Regions for Developing Water-saving Society*, China WaterPower Press, Beijing, 2004, 3-5.

Watkin, T. G. (1999) *An Historical Introduction to Modern Civil Law*, Dartmouth, Aldershot.

WHO (2000), *The Global Water Supply and Sanitation Assessment 2000*, WHO, Geneva.

WHO (2003) *The Right To Water*, WHO.

Wilkinson, D. (2002) *Environment and Law*, Routledge, New York.

World Bank (2003), *Water Resources Sector Strategy: Strategic Directions for World Bank Engagement*, World Bank Publication, Washington, D.C.

Wouters, P. K., et al. (2001) The Legal Response to the World's Crisis: What Legacy from the Hague ? *Water Law Review* **4** (2), 419-425.

Wouters, P. K., et al. (2004) The New Development of Water Law in China, *Water Law Review* **7** (2) 243-307.

Wu, G. (2001) A Letter to the Minister Wang Shucheng, *China Water Resources News*, 28 February 2001.

Xin, C. (1999) *Chinese Legal System & Current Legal Reform*, (China) Law Press, Beijing.

Yan, W. (2001) *Chinese Legal Reform*, Routledge, London.

Ye, Q. (1998) Flow Interruptions and Their Environmental Impacts on the Yellow River Delta, *ACTA GEOGRAPHY SINICA* (1998) 5, 385-391.

Zhu, X. (2001) Adapting New Measures to Strengthen Comprehensive Rehabilitation in Tarim River Basin in the New Situation, *China Water Resources* (2001) 9, 50-52.

Index

IWA Publishing's authorised EU representative for General Product Safety Regulations is Diane D'Arras, 15 rue Duret, 75116 Paris, France, e-mail: safety@iwap.co.uk.

Printed and bound by CPI Group (UK) Ltd, Croydon, CR0 4YY

12/05/2026

02109005-0001